AN

INTELLECTUAL ARITHMETIC,

UPON THE

INDUCTIVE METHOD,

WITH AN

INTRODUCTION TO WRITTEN ARITHMETIC.

BY

JAMES S. EATON, M. A.,

INSTRUCTOR IN PHILLIPS ACADEMY, ANDOVER, AND AUTHOR OF "EASY LESSONS IN MENTAL ARITHMETIC," "THE COMMON SCHOOL ARITHMETIC," AND "A TREATISE ON WRITTEN ARITHMETIC."

BOSTON:
THOMPSON, BIGELOW & BROWN.
29 CORNHILL.
1872.

PREFACE.

THE Pestalozzian or Inductive Method of teaching the science of numbers is now universally approved by intelligent teachers. The first attempt in this country to apply this method to Mental Arithmetic resulted in the publication of Colburn's First Lessons, a work whose success has not exceeded its merit. It was, however, a useful experiment rather than a perfect realization of the inductive system of instruction. That the subsequent books of the same class and purpose have failed to correct its defects, and thus meet the demand it created, is due evidently to their departure from the true theory as developed and exemplified by Pestalozzi.

The Author of this work has endeavored to improve upon all his predecessors, by adhering more closely than even Colburn did to the original method of the great Swiss educator, and by presenting at the same time, in a practical and attractive form, such improvements in the application of his principles as have stood the test of enlightened experience.

In accordance with this design, the subjects are so arranged that each step of the learner prepares him for that which follows. By this suggestive and natural order of arrangement, together with copious illustrations of principles and applications by means of *small concrete numbers,* the pupil is led to a clear apprehension of the properties and relations of numbers, and is enabled to understand everything as he advances, till he acquires a thorough knowledge of the nature and use of the essential numerical operations.

While the general arrangement of the subjects and examples is *strictly progressive and logical*, the difficulty of the problems is occasionally varied, in order to prevent the weariness of a long, unbroken ascent, and to afford a grateful alternation of effort and relaxation, like that experienced by the traveler in crossing a country diversified by hill, valley, and plain.

The analytical process which this method requires at every step is calculated to develop and strengthen the mental powers, and to form the habit of rapid and accurate thought. Some illustrations of modes of analyzing questions have been presented merely as suggestions to the pupil; but the plan of the work does not embrace *set forms of analysis* for the various classes of examples, a contrivance little likely to stimulate invention or promote self-reliance. On the contrary, its distinctive feature is its special adaptation to the mode of teaching which leads the learner to ascertain for himself each step to be taken, to think and reason independently, and to rely upon his own powers and resources, thus securing a vigorous and healthful discipline of his intellectual faculties.

Though this work is intended as a connecting link between the Primary and Written Arithmetics of the Author, thus completing the Series on which he has been so long engaged, it is also complete in itself. It presents a mental analysis of Arithmetic adapted to the younger pupils by its easy gradations, and to advanced pupils by its scientific arrangement and its logical development of the art of computation; and yet it has been limited to the true province of Intellectual Arithmetic, which is to serve as an *introduction* to Written Arithmetic, and not as a substitute for it, as some authors seem to imagine.

In the spirit of the inductive method, concrete numbers—numbers applied to physical objects—have been largely employed in treating of each topic, as the only fit preparation for the exercises upon abstract numbers, which are far more difficult for the youthful mind to grasp.

A few pages of Written Arithmetic have been appended, embracing examples in the ground rules and compound numbers, which may be profitably studied in connection with the mental lessons illustrating the same principles.

Fully aware of the difficulty of the task he has undertaken, the Author has spared no pains in its execution, and he gratefully acknowledges his obligations for the numerous valuable suggestions with which he has been favored by several eminent practical teachers.

The favorable reception of the other books of his Series, encourages him to hope that this attempt to *perfect* and *modernize* the *original Inductive System* of *Mental Arithmetic*, and adapt it to the wants of schools of the present day, will meet with the general approbation of teachers and educators.

PHILLIPS ACADEMY, ANDOVER,
MARCH 14, 1864.

SUGGESTIONS TO TEACHERS.

THE teacher who would attain high success must *study methods*, and never take it for granted that he is perfect in his art. Why does one teacher accomplish twice as much as another, with no greater expenditure of time and strength? Because he has twice as much *skill*. Skill is *acquired*. It is gained by experimenting; that is, by experience guided by good judgment, and enlightened by the study of methods and expedients. The following Suggestions, derived from long experience and much study of the subject of teaching Mental Arithmetic, are submitted for your consideration, and not as rules which you are to blindly follow without the exercise of independent thought.

1. Take great pains in *assigning* the lesson, adapting its length to the capacity of the class, stating explicitly how it is to be learned and in what manner it is to be recited, and giving sufficient time for its thorough preparation.

2. See that the lesson is faithfully *studied*. Many teachers waste time over lessons which have not been properly prepared. Sometimes study a lesson with the pupils, to show them how.

3. Do *not* require the pupils to *commit the questions to memory*. This is a waste of time. Nor should they commit the answers, *excepting* the answers to that class of examples which involve a single operation upon abstract numbers; that is, such questions as are usually comprised in the tables of addition, subtraction, multiplication, and division.

4. Never require a pupil to analyze questions according to a *set form of analysis*, but encourage originality in methods of solution. The fewer words in the solution the better, if it is correct and intelligible. By all means avoid long and complicated formulas.

5. Do not demand reasons for answers which require no process of analysis. If the child knows that 4 from 6 leaves 2, what is gained by requiring him to say, *Because* 4 and 2 are 6? The thing is no better understood, and time is consumed.

6. The teacher will read the questions himself, the class dispensing with the book, or he will allow the pupils to have the book and read the examples, as he may prefer. In questions requiring analysis the pupils should not be called in turn, but *promiscuously* or *by cards*, and, if the example is read by the teacher, time should be given, after the reading, for the class *to think*, before any pupil is designated to answer. Examples like those in Lesson II, page 11, may be recited by the members of the class *in rotation*, the questions being read *rapidly*.

7. The answer to a question requiring a process of solution should not be given before the solution, but it should be given at the conclusion of the solution. Nor should pupils be required, as a practice, to give what may be called an abstract or general answer before the solution, like the following: $\frac{4}{9}$ of 36 is $\frac{4}{9}$ of

how many times $\frac{2}{7}$ of 42? As many times $\frac{2}{7}$ of 42 as $\frac{2}{7}$ of 42 is contained times in the number of which $\frac{1}{4}$ of $\frac{4}{3}$ of 36 is $\frac{1}{9}$. Such exercises may be good discipline, but there is no need of consuming time on exercises merely for discipline, as the opportunities for it in acquiring useful knowledge are abundant.

8. When practicable, it is best that the whole class should stand at recitation. At any rate, the pupil who recites should stand, and, if the teacher reads the example, the pupil should repeat it after him, before giving the explanation.

9. Never proceed with the recitation unless *every member* of the class is giving attention, but do not try to keep the attention too long. Many expedients must be employed to keep the attention awake. Sometimes the pupils may "take places," sometimes they may be permitted to correct each other, and sometimes a pupil may be called at random to finish a solution commenced by another.

10. Aim at *thoroughness* in every step. This is much promoted by frequent and judicious *reviews*. With every lesson in advance the preceding should be reviewed; and there should be monthly and quarterly reviews beside.

11. If you suspect that a solution has been committed to memory without being understood, give a similar original question with different numbers.

12. Where it is practicable, illustrate problems and principles by sensible objects. Let fractions be illustrated by dividing an apple, a line, a square, or some other object. The tables of weights and measures should be taught according to the method of object-teaching, and not abstractly committed to memory.

13. As an occasional exercise, let each pupil, from memory, propose to the pupil next above him some question embraced in the part of the book which has been studied, the pupil failing to solve the question put to him losing his place; or, where "place taking" is not practiced, let there be a forfeiture of merits for failure, or a gain for success.

14. Original questions similar to examples 32 and 33, page 28, to be answered simultaneously by the class, should be proposed frequently and enunciated rapidly.

15. The learner should seldom if ever be told directly how to perform any operation in Arithmetic. Much less should he have the operation performed for him. Instead of telling the pupil directly how to go on, examine him, and endeavor to discover in what his difficulty consists, and then, if possible, remove it.

16. The recitation should be conducted *briskly*, and it should be so managed, if practicable, that each pupil shall endeavor to solve every question proposed; but it is not necessary that the whole lesson should be actually *recited* by each pupil.

17. But the most important requisite to success is *to create and to sustain an interest in the study*. How can this be done? In the first place you must be really *very much interested yourself*. In the second place, you must *teach well*. And if you are deeply interested in the subject, you will be very likely to find out how to teach it skillfully.

CONTENTS.

INTELLECTUAL ARITHMETIC.

SECTION FIRST.

LESSON I.

1. HENRY had one knife, and he has found another; how many knives has he now?

2. Charles bought an orange for two cents, and an apple for one cent; how many cents did he pay for both?

3. Mary gave two peaches to Sarah, and kept two herself; how many peaches had she at first?

4. If you have four marbles in one hand, and three in the other, how many have you in both?

5. If you have four fingers on each hand, how many fingers have you on both hands?

6. James found five apples under one tree, and four under another; how many apples did he find under both trees?

7. Addie has five canaries, and Ella has three; how many canaries have they together?

8. Robert had three peaches, but he has given one of them away; how many peaches has he now?

9. John having four cents, spent two of them for an orange; how many cents has he now?

10. David has five cents in one hand, and two cents in the other; how many cents has he in both hands? How many more in one hand than in the other?

11. Frank has six gray squirrels, and Herbert has three; how many squirrels have they both?

12. Edwin has six figs, and Philip has four; how many more figs has Edwin than Philip? How many have they both?

13. William having eight plums, gave five of them to Louisa; how many did he keep? How many less did he keep than he gave away?

14. Edward has six doves, and Charles has eight; how many more has Charles than Edward? How many have they both?

15. Lewis bought a pig for eight dollars, and sold it for ten dollars; how much did he gain?

16. A man bought a sled for ten dollars, and sold it for seven; how much did he lose?

17. A man owing ten dollars, paid four dollars; how much did he still owe?

18. A man owing ten dollars, paid all but four dollars; how much did he pay?

19. Mr. Adams sold a pig for three dollars, and a sheep for six dollars; how many dollars did he receive for both?

20. Albert has eight rabbits, and Arthur has two; how many rabbits have they both?

21. George found seven eggs in one nest, and three in another; how many eggs did he find?

22. Frank gave three cents for an orange, and had seven cents left; how many cents had he at first?

23. If a melon is worth ten cents, and an orange is worth four cents, how many cents are they both worth? How much more is the melon worth than the orange?

24. Mary had nine cents, and her mother gave her three; how many cents had Mary then?

25. Robert had ten peaches, but he has given four of them to David; how many peaches has Robert now? How many more than David?

26. Mr. Day bought a barrel of flour for ten dollars, and sold it for two dollars more than he gave for it; how much did he receive for it?

27. A boy having twelve walnuts, gave away five of them; how many had he left?

28. Bought a ton of coal for ten dollars, and a cord of wood for six dollars; what did I pay for both? How much more for the coal than for the wood?

LESSON II.

1. Two and one are how many?
2. Three and one are how many?
3. Four and one are how many?
4. Five and one are how many?
5. Six and one are how many?
6. Seven and one are how many?
7. Eight and one are how many?
8. Nine and one are how many?
9. Ten and one are how many?
10. Two and two are how many?
11. Three and two are how many?
12. Four and two are how many?
13. Five and two are how many?
14. Six and two are how many?
15. Seven and two are how many?
16. Eight and two are how many?
17. Nine and two are how many?
18. Ten and two are how many?
19. Two and three are how many?
20. Three and three are how many?
21. Four and three are how many?
22. Five and three are how many?
23. Six and three are how many?
24. Seven and three are how many?
25. Eight and three are how many?
26. Nine and three are how many?
27. Ten and three are how many?
28. Two and four are how many?
29. Three and four are how many?

30. Four and four are how many?
31. Five and four are how many?
32. Six and four are how many?
33. Seven and four are how many?
34. Eight and four are how many?
35. Nine and four are how many?
36. Ten and four are how many?
37. Two and five are how many?
38. Three and five are how many?
39. Four and five are how many?
40. Five and five are how many?
41. Six and five are how many?
42. Seven and five are how many?
43. Eight and five are how many?
44. Nine and five are how many?
45. Ten and five are how many?
46. Two and six are how many?
47. Three and six are how many?
48. Four and six are how many?
49. Five and six are how many?
50. Six and six are how many?
51. Seven and six are how many?
52. Eight and six are how many?
53. Nine and six are how many?
54. Ten and six are how many?
55. Two and seven are how many?
56. Three and seven are how many?
57. Four and seven are how many?
58. Five and seven are how many?
59. Six and seven are how many?
60. Seven and seven are how many?
61. Eight and seven are how many?
62. Nine and seven are how many?
63. Ten and seven are how many?
64. Two and eight are how many?
65. Three and eight are how many?
66. Four and eight are how many?

67. Five and eight are how many?
68. Six and eight are how many?
69. Seven and eight are how many?
70. Eight and eight are how many?
71. Nine and eight are how many?
72. Ten and eight are how many?
73. Two and nine are how many?
74. Three and nine are how many?
75. Four and nine are how many?
76. Five and nine are how many?
77 Six and nine are how many?
78. Seven and nine are how many?
79. Eight and nine are how many?
80. Nine and nine are how many?
81. Ten and nine are how many?
82. Two and ten are how many?
83. Three and ten are how many?
84. Four and ten are how many?
85. Five and ten are how many?
86. Six and ten are how many?
87. Seven and ten are how many?
88. Eight and ten are how many?
89. Nine and ten are how many?
90. Ten and ten are how many?

LESSON III.

1. JOSEPH bought three apples for three cents, and seven apples for seven cents; how many apples did he buy? How many cents did he pay for all the apples?

2. John bought three tops for six cents, and Abel bought nine tops for eight cents; how many tops did they both buy? How many cents did they pay for them?

3. A man bought three barrels of apples for eight dollars, and twelve bushels of potatoes for nine dollars;

how much did he give for the whole? How much less for the apples than for the potatoes?

4. A man owing fifteen dollars, paid five dollars; how much did he still owe?

5. A man owing fourteen dollars, paid all but six dollars; how much did he pay?

6. A lady bought ten yards of ribbon, and nine yards of braid; how many yards of both did she buy? How much more ribbon than braid?

7. A man bought three sheep for fifteen dollars, and six lambs for five dollars; how many animals did he buy? How many more lambs than sheep? How many dollars did he pay for all?

8. A boy bought sixteen marbles, but lost five of them; how many had he left?

9. A boy having eighteen marbles, lost a part of them, and had twelve left; how many did he lose?

10. A man bought a cow for twenty dollars, but could not sell her for so much by six dollars; for what sum could he sell her?

11. Two boys, Richard and Martin, played at marbles. When they began to play they had eight marbles apiece, but when they finished their game, Richard had won three; how many marbles had each then?

12. A man bought eighteen pounds of sugar, and lost six pounds of it as he was carrying it home; how many pounds had he left?

13. A boy bought nineteen chickens, and a cat killed all but ten of them; how many did the cat kill?

14. A merchant bought a firkin of butter for eighteen dollars, but, it being damaged, he sold it again for fourteen dollars; how much did he lose?

15. A farmer bought a colt for fifteen dollars; he paid five dollars for keeping it, and then sold it for twenty-three dollars; how much did he gain?

16. Ten boys and nine boys are how many boys?

17. Ten birds and nine birds are how many birds?

LESSON IV.

1. Two and one are how many?
2. Two and two are how many?
3. Four and two are how many?
4. Seven and two are how many?
5. Five and two are how many?
6. Nine and two are how many?
7. Six and two are how many?
8. Eight and two are how many?
9. Three and two are how many?
10. Ten and two are how many?
11. Two and three are how many?
12. Six and three are how many?
13. Four and three are how many?
14. Seven and three are how many?
15. Three and three are how many?
16. Five and three are how many?
17. Eight and three are how many?
18. Ten and three are how many?
19. Nine and three are how many?
20. Five and four are how many?
21. Two and four are how many?
22. Four and four are how many?
23. Three and four are how many?
24. Six and four are how many?
25. Nine and four are how many?
26. Seven and four are how many?
27. Ten and four are how many?
28. Eight and four are how many?
29. Two and five are how many?
30. Three and six are how many?
31. Six and five are how many?
32. Seven and six are how many?
33. Nine and six are how many?
34. Eight and five are how many?
35. Five and five are how many?

36. Four and six are how many?
37. Six and six are how many?
38. Nine and five are how many?
39. Eight and six are how many?
40. Six and ten are how many?
41. Seven and eight are how many?
42. Eight and nine are how many?
43. Ten and eight are how many?
44. Eight and seven are how many?
45. Nine and nine are how many?
46. Ten and five are how many?
47. Ten and seven are how many?
48. Nine and eight are how many?
49. Ten and ten are how many?
50. Three and ten are how many?
51. Seven and seven are how many?
52. Five and seven are how many?
53. Nine and seven are how many?
54. Eight and eight are how many?
55. Three and eight are how many?
56. Seven and ten are how many?
57. Two and ten are how many?
58. Four and ten are how many?
59. Two and eight are how many?
60. Four and seven are how many?
61. Four and eight are how many?
62. Nine and ten are how many?

LESSON V.

1. THREE less two are how many?
2. Five less one are how many?
3. Six less three are how many?
4. Six less two are how many?
5. Seven less five are how many?
6. Six less four are how many?
7. Eight less three are how many?

8. Eight less five are how many?
9. Nine less one are how many?
10. Nine less five are how many?
11. Five less three are how many?
12. Nine less three are how many?
13. Ten less two are how many?
14. Eight less six are how many?
15. Eight less two are how many?
16. Ten less seven are how many?
17. Eleven less ten are how many?
18. Eleven less six are how many?
19. Ten less five are how many?
20. Ten less nine are how many?
21. Twelve less six are how many?
22. Twelve less nine are how many?
23. Eight less four are how many?
24. Thirteen less six are how many?
25. Thirteen less two are how many?
26. Fifteen less ten are how many?
27. Twenty less ten are how many?
28. Sixteen less six are how many?
29. Sixteen less ten are how many?
30. Fifteen less four are how many?
31. Eighteen less eight are how many?
32. Eighteen less ten are how many?
33. Nineteen less one are how many?
34. Seventeen less three are how many?
35. Seventeen less ten are how many?
36. Eighteen less six are how many?
37. Nineteen less nine are how many?
38. Twenty less five are how many?
39. Twenty-five less five are how many?
40. Twenty-six less three are how many?
41. Eight and six, less four, are how many?
42. Ten and eight, less six, are how many?
43. Four and nine, less seven, are how many?
44. Sixteen and six, less two, are how many?

LESSON VI.

1. THREE boys, James, William, and George, went a fishing. James caught six fishes, William caught four, and George caught three; how many did they all catch? How many more did James catch than George?

2. Three girls, Mary, Ella, and Frances, gave some flowers to their teacher. Mary gave her seven; Ella, three; and Frances, four: how many flowers did the teacher receive? How many less did Ella give than Mary?

3. Joseph had seven peaches; David, seven; and Thomas, nine. Joseph gave two to George, David gave him three, and Thomas gave him one; how many peaches did each of the boys then have?

4. A boy gave five apples to one of his companions; to another, eight; to another, three; and kept four himself: how many had he at first?

5. A generous boy having eighteen peaches, gave three of them to one companion, six to another, and five to another; how many did he give away? How many did he keep?

6. A man paid three dollars for some sugar, six dollars for two barrels of apples, ten dollars for a barrel of flour, and had six dollars left; how many dollars had he at first?

7. A boy having twenty-five cents, paid ten cents for a melon, four cents for an orange, and gave away six cents; how many cents had he remaining?

8. A boy bought a ball for sixteen cents; he gave ten cents to have it covered, and then sold it for thirty-two cents; how much did he gain by the bargains?

9. A man bought a cow for eighteen dollars; he paid nine dollars for pasturing her, and then sold her for twenty-five dollars; how much did he lose by the bargains?

10. A tailor bought some cloth for fifteen dollars; he paid seven dollars to have it made into a coat, and then sold the coat for twenty-seven dollars; how much did he gain on the coat?

LESSON VII.

1. If one lemon is worth two cents, what are two lemons worth? What are three worth?
2. What cost two oranges, at three cents apiece?
3. What cost four yards of cloth, at three dollars a yard? What at four dollars a yard?
4. There are three feet in one yard; how many feet are there in five yards? In six yards? In six yards and one foot? In ten yards and two feet?
5. Bought two oranges at five cents apiece, and four lemons at three cents apiece; what did I pay for the oranges? What for the lemons? For both?
6. Sold seven barrels of apples at two dollars a barrel, and two cords of wood at five dollars a cord; how many dollars did I receive for the apples? For the wood? For both?
7. What cost five yards of ribbon, at ten cents a yard? At nine cents?
8. If I recite three lessons each day, how many shall I recite in six days? In ten days?
9. If four bushels of wheat make a barrel of flour, how many bushels will make six barrels? Eight barrels?
10. What cost five pineapples, at ten cents each? At eight cents?
11. What will six barrels of flour cost, at eight dollars a barrel? At ten dollars?
12. If a horse travel eight miles in an hour, how far will he travel in seven hours?
13. What cost ten pounds of sugar, at nine cents a pound? At ten cents?

14. What is the value of nine doves, at nine cents apiece? At eleven cents?

15. If a man earn seven dollars in one week, how much will he earn in seven weeks? In ten weeks?

LESSON VIII.

1. Two times one are how many?
2. Two times two are how many?
3. Two times three are how many?
4. Two times four are how many?
5. Two times five are how many?
6. Two times six are how many?
7. Two times seven are how many?
8. Two times eight are how many?
9. Two times nine are how many?
10. Two times ten are how many?
11. Two times eleven are how many?
12. Two times twelve are how many?
13. Three times one are how many?
14. Three times two are how many?
15. Three times three are how many?
16. Three times four are how many?
17. Three times five are how many?
18. Three times six are how many?
19. Three times seven are how many?
20. Three times eight are how many?
21. Three times nine are how many?
22. Three times ten are how many?
23. Three times eleven are how many?
24. Three times twelve are how many?
25. Four times one are how many?
26. Four times two are how many?
27. Four times three are how many?
28. Four times four are how many?
29. Four times five are how many?
30. Four times six are how many?

31. Four times seven are how many?
32. Four times eight are how many?
33. Four times nine are how many?
34. Four times ten are how many?
35. Four times eleven are how many?
36. Four times twelve are how many?
37. Five times one are how many?
38. Five times two are how many?
39. Five times three are how many?
40. Five times four are how many?
41. Five times five are how many?
42. Five times six are how many?
43. Five times seven are how many?
44. Five times eight are how many?
45. Five times nine are how many?
46. Five times ten are how many?
47. Five times eleven are how many?
48. Five times twelve are how many?
49. Six times one are how many?
50. Six times two are how many?
51. Six times three are how many?
52. Six times four are how many?
53. Six times five are how many?
54. Six times six are how many?
55. Six times seven are how many?
56. Six times eight are how many?
57. Six times nine are how many?
58. Six times ten are how many?
59. Six times eleven are how many?
60. Six times twelve are how many?
61. Count to fifty, by twos, using the even numbers; thus, two, four, six, eight, etc.
62. Count backward, by twos, from fifty; thus, fifty, forty-eight, forty-six, etc.
63. Count by twos to forty-nine, using the odd numbers; thus, one, three, five, etc.
64. Count backward, by twos, from forty-nine; thus, forty-nine, forty-seven, forty-five, etc.

LESSON IX.

1. Two and three, less one, are how many?
2. Five and four, less three, are how many?
3. Ten and two and one, less five, are how many?
4. Six and ten and two, less eight, are how many?
5. Three times six, less four, are how many?
6. Four times five, less nine, are how many?
7. Eight times two, less ten, are how many?
8. Three and six, less four, are how many?
9. Five times five, less five, are how many?
10. Six times six, less six, are how many?
11. Six times six, and six, are how many?
12. Ten times two, less six, are how many?
13. Five times six, and seven, are how many?
14. Eight and six and two, less ten, are how many?
15. Ten and six and five, less nine, are how many?
16. Ten times two, less eight, are how many?
17. Ten and five, less nine, are how many?
18. Ten and eight, less six, are how many?
19. Ten and ten, less five, are how many?
20. Sold a sheep for six dollars and a shote for ten dollars, and afterward paid three dollars for a pig; how many dollars had I then?
21. Nellie found six peaches on one tree and eight on another, but she has eaten two and given three to Jane; how many has she remaining?
22. A boy walked from home twelve miles on one day, and ten miles on the next day, but on the third day he returned seven miles; how far from home was he then?
23. Samuel had forty cents, but he has given five of them for a ball, and three for an orange; how many cents has he now?
24. Daniel sold six doves at twelve cents apiece, and bought five chickens at ten cents apiece; how much money had he left?

25. Thomas sold five melons, at ten cents each, and paid twenty-five cents for a book, and ten cents for a slate; how many cents had he remaining?

26. A man having fifty dollars, paid twenty dollars for a coat, six dollars for a pair of pantaloons, and four dollars for a vest; how many dollars had he left?

27. A boy found fourteen eggs in one nest, and eight less in another; how many eggs did he find in the second nest? How many in both nests?

LESSON X.

1. IF two lambs cost six dollars, what will one lamb cost?

2. How many peaches, at two cents apiece, can you buy for six cents?

3. If I walk four miles per hour, in how many hours shall I walk eight miles?

4. If I walk eight miles in two hours, how many miles do I walk in one hour? How many miles in three hours?

5. At three dollars a yard, how many yards of cloth can I buy for nine dollars? For twelve dollars?

6. If you pay twelve dollars for four yards of cloth, what is the price of one yard?

7. Twelve apples were divided equally between three boys; how many apples did each boy receive?

8. In a certain orchard there are fifteen trees standing in rows, and there are five trees in each row; how many rows are there?

9. I have twenty trees, standing in four equal rows; how many trees are there in each row?

10. How many barrels of apples, at three dollars a barrel, can I buy for fifteen dollars? For eighteen dollars?

11. How many pears, at two cents apiece, can you buy for eighteen cents? For twenty cents?

12. How many oranges, at four cents apiece, can you buy for twenty cents? For twenty-four cents?

13. Bought four barrels of flour for twenty-eight dollars; what was the price of one barrel?

14. Paid twenty-five cents for five oranges; how many cents did I pay for one orange? For two?

15. How many spools of thread, at eight cents a spool, can be bought for thirty-two cents? For forty cents?

16. Paid thirty-six cents for six spools of thread; how many cents did I pay for one spool? For four?

17. A farmer sowed twenty-eight bushels of oats on seven acres of land; how many bushels did he sow on one acre? On five acres?

18. If I pay thirty-five dollars for five cords of wood, what is the price of one cord?

19. If coal is worth six dollars a ton, how many tons can I buy for forty-two dollars? For twenty-four dollars?

20. If a bird can fly twelve miles in one hour, in what time can it fly twenty-four miles? Thirty-six miles?

21. If wood is worth six dollars a cord, and coal is worth eight dollars a ton, how many cords of wood will pay for three tons of coal?

LESSON XI.

1. FOUR are how many times two?
2. Six are how many times two?
3. Six are how many times three?
4. Eight are how many times two? Four?
5. Ten are how many times five? Two?
6. Nine are how many times three?
7. Twelve are how many times three? Six?
8. Twelve are how many times two? Four?
9. Fourteen are how many times seven?

10. Fifteen are how many times five? Three?
11. Sixteen are how many times four? Two?
12. Eighteen are how many times two? Six?
13. Eighteen are how many times three? Nine?
14. Twenty are how many times five? Two?
15. Twenty-one are how many times seven?
16. Twenty-four are how many times six? Two?
17. Twenty-four are how many times four? Eight?
18. Twenty-four are how many times three?
19. Three times four are how many times three?
20. Five times eight are how many times four?
21. Six times five are how many times ten?
22. Four times nine are how many times six?
23. Four times six are how many times eight?
24. Ten times four are how many times eight?
25. Eight times six are how many times twelve?
26. Six in twelve, how many times?
27. Three in twenty-one, how many times?
28. Five in thirty-five, how many times?
29. Eight in thirty-two, how many times?
30. In forty-two, how many times six?
31. In forty-eight, how many times eight?
32. In thirty-five, how many times seven?
33. In forty-nine, how many times seven?
34. In fifty, how many times ten? Five?
35. Eleven in fifty-five, how many times?
36. Eight in fifty-six, how many times?
37. Twelve in sixty, how many times?
38. Eight in seventy-two, how many times?
39. Nine in sixty-three, how many times?
40. Five in fifty-five, how many times?
41. Eight times nine are how many times twelve?
42. Five times twelve are how many times six?
43. Three times fourteen are how many times six?
44. Three times fifteen are how many times nine?
45. Three times twenty are how many times twelve?
46. Eight times nine are how many times six?

SECTION SECOND.

LESSON I.

Remarks. 1. Instead of writing the names of numbers, it is customary to express them by certain marks, or characters, called *figures*.

2. These figures are ten in number, and are called *Arabic* figures.

3. Taken separately, each figure means the same as the word placed under it in the following lines; thus,

0,	1,	2,	3,	4,	5,	6,	7,	8,	9.
Naught,	One,	Two,	Three,	Four,	Five,	Six,	Seven,	Eight,	Nine.

4. To express numbers greater than nine, these figures are repeated and combined in various ways; thus,

10,	ten.	25,	twenty-five.
11,	eleven.	26,	twenty-six.
12,	twelve.	27,	twenty-seven.
13,	thirteen.	28,	twenty-eight.
14,	fourteen.	29,	twenty-nine.
15,	fifteen.	30,	thirty.
16,	sixteen.	31,	thirty-one.
17,	seventeen.	40,	forty.
18,	eighteen.	41,	forty-one.
19,	nineteen.	90,	ninety.
20,	twenty.	99,	ninety-nine.
21,	twenty-one.	100,	one hundred.
22,	twenty-two.	102,	one hundred and two.
23,	twenty-three.	205,	two hundred and five.
24,	twenty-four.	900,	nine hundred.

Note. It is presumed that the pupil is familiar with all the figures, and the manner of combining them so as to express small numbers, like those in the above Table, previous to studying this book. If, however, this is found to be untrue in any case, such pupil should be exercised by the teacher, upon the blackboard or elsewhere, until he can readily write in figures any number less than 1000.

5. To abbreviate expressions, and to indicate the relations of numbers to each other, we sometimes use *signs* instead of words.

6. This mark, \$, is often used as a sign of the word *dollar*, or *dollars;* thus, the expression \$1, stands for *one dollar;* \$4 stands for *four dollars.*

7. The *sign of equality*, $=$, signifies that the quantities between which it stands are equal to each other; thus, \$1 $=$ 100 cents; that is, one dollar equals one hundred cents.

8. The *sign of addition*, $+$, called *plus* or *and*, denotes that the quantities between which it stands are to be added together; thus, $3+2=5$; that is, three plus two equals five, or three and two are five.

9. The *sign of subtraction*, —, called *minus* or *less*, signifies that the number after it is to be taken from the number before it; thus, $7-4=3$; that is, seven minus four, or seven less four, equals three.

10. The *sign of multiplication*, $\times$, signifies that the two numbers between which it stands are to be multiplied together; thus, $6\times 5=30$; that is, six multiplied by five equals thirty, or six times five are thirty.

11. The *sign of division*, $\div$, indicates that the number before it is to be divided by the number after it; thus, $8\div 2=4$; that is, eight divided by two equals four, or two in eight, four times.

LESSON II.

1. Massachusetts has 14 counties, Rhode Island has 5, and Connecticut has 8; how many counties have the three States? *Ans.* $14+5+8=27$.

2. How many more counties has Massachusetts than Rhode Island? *Ans.* $14-5=9$.

3. If I pay \$9 for a barrel of flour, \$8 for a hundred pounds of pork, and \$3 for a bushel of beans, how many dollars shall I pay for all?

4. If a ton of coal costs \$8, and a cord of wood costs \$5, what do they both cost? How much more does the coal cost than the wood?

5. How many are $9 + 8 + 3$? $9 - 3$?

6. How many are $6 + 4 + 7$? $8 - 5$?

7. How many are $7 + 5 + 9$? $9 - 7$?

8. How many are $5 + 8 + 7$? $7 - 5$?

9. How many are $8 + 6 + 7$? $8 - 6$?

10. How many are $6 + 7 + 4$? $9 - 5$?

11. If 1 barrel of flour costs \$9, what will 8 barrels cost? *Ans.* $\$9 \times 8 = \$72$.

12. How many are 12×5? 11×7?

13. How many are 6×7? 9×6?

14. How many are 10×5? 8×7?

15. How many are 5×11? 6×12?

16. How many are 9×7? 5×11?

17. How many are 7×10? 6×8?

18. How many are 9×10? 6×7?

19. How many are 4×12? 7×8?

20. If 8 pounds of sugar cost 72 cents, what is the price of 1 pound? *Ans.* 72 cents $\div 8 = 9$ cents.

21. How many are $56 \div 7$? $63 \div 9$?

22. How many are $32 \div 8$? $24 \div 6$?

23. How many are $40 \div 5$? $42 \div 6$?

24. How many are $56 \div 8$? $56 \div 7$?

25. How many are $60 \div 10$? $45 \div 9$?

26. How many are $72 \div 12$? $66 \div 6$?

27. How many are $49 \div 7$? $64 \div 8$?

28. How many are $8 - 6$? $9 + 7$?

29. How many are $7 \times 8 - 6$? $8 \times 4 + 8$?

30. How many are $48 \div 4 + 3$? $60 \div 10 - 2$?

31. How many are $9 + 6 + 8 - 3$? $48 \div 6 + 3$?

32. Take 6, add 4, multiply by 2, divide by 5, divide by 2, add 7, divide by 3, multiply by 4, subtract 2; result?

33. Take 9, subtract 3, divide by 2, add 5, multiply by 4, subtract 2, divide by 6, add 7; result?

LESSON III.

1. A MAN bought a tub of butter for 12 dollars, and a cheese for 4 dollars ; how many dollars did he pay for both ?

2. A farmer has 8 cows in one pasture, and 12 in another; how many has he in both? How many more in one than in the other ?

3. George found 6 marbles, Arthur gave him 10, and he bought 8 ; how many marbles had he then ?

4. One morning Mary found in her flower-garden 8 roses, 5 peonies, 7 tulips, and 10 other blossoms; she picked 4 roses, 1 peony, 3 tulips, and 7 other blossoms ; how many blossoms did she find? How many did she leave in the garden ?

5. A farmer sold 8 bushels of wheat, 9 bushels of corn, 3 bushels of rye, and 7 bushels of barley ; how many bushels of grain did he sell?

6. A drover bought 10 sheep of one man, 7 of another, and 8 of another, and he afterwards sold 12 sheep; how many did he buy? How many did he keep?

7. If I earn 12 dollars, and spend 7 dollars, a week, how many dollars shall I save in a week? How many in 4 weeks ?

8. Bought 9 yards of cloth, at 4 dollars a yard, and sold the whole of it for 9 dollars more than I paid; what did I receive for the 9 yards ? What for each yard ?

9. Bought 8 yards of cloth, at 5 dollars a yard, and sold all of it for 32 dollars; did I gain or lose ? How much on the whole ? How much on a yard?

10. Mr. Flint had 10 sheep in each of 5 pastures, but he has sold 5 from one pasture, 7 from another, and 8 from another; how many sheep has he now ?

11. When flour is worth 8 dollars a barrel, and hay 12 dollars a ton, how many tons of hay will pay for 6 barrels of flour?

12. In a certain orchard there were 7 rows of trees, and 8 trees in each row, but 2 trees have been taken from each row ; how many trees still remain in the orchard ?

13. Daniel, Joseph, and James went a fishing. James caught 2 fishes, Joseph caught 3 times as many as James, and Daniel caught twice as many as James and Joseph together ; how many did Daniel catch ?

14. Thomas had twice 15 peaches, which he divided equally between his 3 brothers ; how many did he give to each ?

15. The distance from Boston to Malden is 5 miles ; from Malden to Reading, 7 miles ; from Reading to Wilmington, 3 miles ; from Wilmington to Andover, 8 miles ; from Andover to Lawrence, 3 miles : how far is it from Boston to Lawrence ?

16. The distance from Boston to Somerville is 3 miles ; from Somerville to Waltham, 7 miles ; from Waltham to Concord, 10 miles ; from Concord to Littleton, 11 miles ; from Littleton to Groton, 4 miles ; from Groton to Fitchburg, 15 miles : how far is it from Boston to Fitchburg ?

LESSON IV.

THE following exercises are designed to discipline the pupil in changing from one subject to another with rapidity and accuracy.

Let the examples be read *across* the page ; thus, 1 and 1 are 2 ; 1 from 2 leaves 1 ; once 1 is 1 ; 1 in 1, once ; etc. etc.

1 and 1 ?	1 from 2 ?	once 1 ?	1 in 1 ?
1 and 2 ?	1 from 3 ?	once 2 ?	1 in 2 ?
1 and 3 ?	1 from 4 ?	once 3 ?	1 in 3 ?
1 and 4 ?	1 from 5 ?	once 4 ?	1 in 4 ?
1 and 5 ?	1 from 6 ?	once 5 ?	1 in 5 ?
1 and 6 ?	1 from 7 ?	once 6 ?	1 in 6 ?

1 and 7?	1 from 8?	once 7?	1 in 7?
1 and 8?	1 from 9?	once 8?	1 in 8?
1 and 9?	1 from 10?	once 9?	1 in 9?
1 and 10?	1 from 11?	once 10?	1 in 10?
1 and 11?	1 from 12?	once 11?	1 in 11?
1 and 12?	1 from 13?	once 12?	1 in 12?
2 and 1?	2 from 3?	2 times 1?	2 in 2?
2 and 2?	2 from 4?	2 times 2?	2 in 4?
2 and 3?	2 from 5?	2 times 3?	2 in 6?
2 and 4?	2 from 6?	2 times 4?	2 in 8?
2 and 5?	2 from 7?	2 times 5?	2 in 10?
2 and 6?	2 from 8?	2 times 6?	2 in 12?
2 and 7?	2 from 9?	2 times 7?	2 in 14?
2 and 8?	2 from 10?	2 times 8?	2 in 16?
2 and 9?	2 from 11?	2 times 9?	2 in 18?
2 and 10?	2 from 12?	2 times 10?	2 in 20?
2 and 11?	2 from 13?	2 times 11?	2 in 22?
2 and 12?	2 from 14?	2 times 12?	2 in 24?
3 and 1?	3 from 4?	3 times 1?	3 in 3?
3 and 2?	3 from 5?	3 times 2?	3 in 6?
3 and 3?	3 from 6?	3 times 3?	3 in 9?
3 and 4?	3 from 7?	3 times 4?	3 in 12?
3 and 5?	3 from 8?	3 times 5?	3 in 15?
3 and 6?	3 from 9?	3 times 6?	3 in 18?
3 and 7?	3 from 10?	3 times 7?	3 in 21?
3 and 8?	3 from 11?	3 times 8?	3 in 24?
3 and 9?	3 from 12?	3 times 9?	3 in 27?
3 and 10?	3 from 13?	3 times 10?	3 in 30?
3 and 11?	3 from 14?	3 times 11?	3 in 33?
3 and 12?	3 from 15?	3 times 12?	3 in 36?
4 and 1?	4 from 5?	4 times 1?	4 in 4?
4 and 2?	4 from 6?	4 times 2?	4 in 8?
4 and 3?	4 from 7?	4 times 3?	4 in 12?
4 and 4?	4 from 8?	4 times 4?	4 in 16?
4 and 5?	4 from 9?	4 times 5?	4 in 20?
4 and 6?	4 from 10?	4 times 6?	4 in 24?

4 and 7?	4 from 11?	4 times 7?	4 in 28?
4 and 8?	4 from 12?	4 times 8?	4 in 32?
4 and 9?	4 from 13?	4 times 9?	4 in 36?
4 and 10?	4 from 14?	4 times 10?	4 in 40?
4 and 11?	4 from 15?	4 times 11?	4 in 44?
4 and 12?	4 from 16?	4 times 12?	4 in 48?

LESSON V.

1. How many are nine and two? Nineteen and two? Twenty-nine and two? Thirty-nine and two? Forty-nine and two? Fifty-nine and two? Sixty-nine and two? Seventy-nine and two? Eighty-nine and two? Ninety-nine and two?

2. How many are nine and three? Nineteen and three? Twenty-nine and three? Thirty-nine and three? Forty-nine and three? Fifty-nine and three? Sixty-nine and three? Seventy-nine and three? Eighty-nine and three? Ninety-nine and three?

3. How many are nine and four? Nineteen and four? Twenty-nine and four? Thirty-nine and four? Forty-nine and four? Fifty-nine and four? Sixty-nine and four? Seventy-nine and four? Eighty-nine and four? Ninty-nine and four?

4. How many are nine and five? Nineteen and five? Twenty-nine and five? Thirty-nine and five? Forty-nine and five? Fifty-nine and five? Sixty-nine and five? Seventy-nine and five? Eighty-nine and five? Ninety-nine and five?

5. How many are nine and six? Nineteen and six? Twenty-nine and six? Thirty-nine and six? Forty-nine and six? Fifty-nine and six? Sixty-nine and six? Seventy-nine and six? Eighty-nine and six? Ninety-nine and six?

6. How many are nine and seven? Nineteen and seven? Twenty-nine and seven? Thirty-nine and seven? Forty-nine and seven? Fifty-nine and seven?

Sixty-nine and seven? Seventy-nine and seven? Eighty-nine and seven? Ninety-nine and seven?

7. How many are nine and eight? Nineteen and eight? Twenty-nine and eight? Thirty-nine and eight? Forty-nine and eight? Fifty-nine and eight? Sixty-nine and eight? Seventy-nine and eight? Eighty-nine and eight? Ninety-nine and eight?

8. How many are nine and nine? Nineteen and nine? Twenty-nine and nine? Thirty-nine and nine? Forty-nine and nine? Fifty-nine and nine? Sixty-nine and nine? Seventy-nine and nine? Eighty-nine and nine? Ninety-nine and nine?

9. How many are nine and ten? Nineteen and ten? Twenty-nine and ten? Thirty-nine and ten? Forty-nine and ten? Fifty-nine and ten? Sixty-nine and ten? Seventy-nine and ten? Eighty-nine and ten? Ninety-nine and ten?

10. How many are eight and three? Eighteen and three? Twenty-eight and three? Thirty-eight and three? Forty-eight and three? Fifty-eight and three? Sixty-eight and three? Seventy-eight and three? Eighty-eight and three? Ninety-eight and three?

11. How many are eight and four? Eighteen and four? Twenty-eight and four? Thirty-eight and four? Forty-eight and four? Fifty-eight and four? Sixty-eight and four? Seventy-eight and four? Eighty-eight and four? Ninety-eight and four?

12. How many are eight and five? Eighteen and five? Twenty-eight and five? Thirty-eight and five? Forty-eight and five? Fifty-eight and five? Sixty-eight and five? Seventy-eight and five? Eighty-eight and five? Ninety-eight and five?

13. How many are eight and six? Eighteen and six? Twenty-eight and six? Thirty-eight and six? Forty-eight and six? Fifty-eight and six? Sixty-eight and six? Seventy-eight and six? Eighty-eight and six? Ninety-eight and six?

14. How many are eight and seven? Eighteen and seven? Twenty-eight and seven? Thirty-eight and seven? Forty-eight and seven? Fifty-eight and seven? Sixty-eight and seven? Seventy-eight and seven? Eighty-eight and seven? Ninety-eight and seven?

15. How many are eight and eight? Eighteen and eight? Twenty-eight and eight? Thirty-eight and eight? Forty-eight and eight? Fifty-eight and eight? Sixty-eight and eight? Seventy-eight and eight? Eighty-eight and eight? Ninety-eight and eight?

16. How many are eight and nine? Eighteen and nine? Twenty-eight and nine? Thirty-eight and nine? Forty-eight and nine? Fifty-eight and nine? Sixty-eight and nine? Seventy-eight and nine? Eighty-eight and nine? Ninety-eight and nine?

17. How many are eight and ten? Eighteen and ten? Twenty-eight and ten? Thirty-eight and ten? Forty-eight and ten? Fifty-eight and ten? Sixty-eight and ten? Seventy-eight and ten? Eighty-eight and ten? Ninety-eight and ten?

LESSON VI.

1. A MAN bought a cow for 28 dollars, a sheep for 10 dollars, and a calf for 6 dollars; how much did he pay for all?

2. A farmer sold a tub of butter for 19 dollars, some cheese for 9 dollars, and 3 barrels of apples for 6 dollars; how much did he receive for all?

3. A drover bought 48 sheep of one man, 10 of another, and 7 of another; how many sheep did he buy?

4. A boy gave 39 cherries to one of his companions, 9 to another, and 3 to another; how many cherries did he give away?

5. Bought a piece of land for 79 dollars, another

piece for 10 dollars, and another for 9 dollars; what did I pay for the whole?

6. A grocer bought 4 barrels of flour for 28 dollars, 20 bushels of oats for 10 dollars, and a hundred weight of sugar for 9 dollars; what did he pay for the whole?

7. How many are forty-eight and ten and six?

8. How many are seventy and nine and six?

9. How many are fifty-eight and ten, less six?

10. Mr. Smith bought a horse for 69 dollars, and paid 10 dollars for keeping him and 3 dollars for shoeing him. He let the horse go on a journey, for which he received 12 dollars, and then sold him for 75 dollars; how much did he gain?

11. A man bought a piece of land for 89 dollars, and paid 9 dollars for plowing a part of it, and 10 dollars for fencing it. He received 15 dollars for the grass that grew on it, and then sold it for 100 dollars; how much did he gain?

LESSON VII.

1. How many are 7 and 4? 17 and 4? 27 and 4? 37 and 4? 47 and 4? 57 and 4? 67 and 4? 77 and 4? 87 and 4? 97 and 4?

2. How many are 7 and 5? 17 and 5? 27 and 5? 37 and 5? 47 and 5? 57 and 5? 67 and 5? 77 and 5? 87 and 5? 97 and 5?

3. How many are 7 and 6? 17 and 6? 27 and 6? 37 and 6? 47 and 6? 57 and 6? 67 and 6? 77 and 6? 87 and 6? 97 and 6?

4. How many are 7 and 7? 17 and 7? 27 and 7? 37 and 7? 47 and 7? 57 and 7? 67 and 7? 77 and 7? 87 and 7? 97 and 7?

5. How many are 7 and 8? 17 and 8? 27 and 8? 37 and 8? 47 and 8? 57 and 8? 67 and 8? 77 and 8? 87 and 8? 97 and 8?

6. How many are 7 and 9? 17 and 9? 27 and 9? 37 and 9? 47 and 9? 57 and 9? 67 and 9? 77 and 9? 87 and 9? 97 and 9?

7. How many are 7 and 10? 17 and 10? 27 and 10? 37 and 10? 47 and 10? 57 and 10? 67 and 10? 77 and 10? 87 and 10? 97 and 10?

8. How many are 6 and 5? 16 and 5? 26 and 5? 36 and 5? 46 and 5? 56 and 5? 66 and 5? 76 and 5? 86 and 5? 96 and 5?

9. How many are 6 and 6? 16 and 6? 26 and 6? 36 and 6? 46 and 6? 56 and 6? 66 and 6? 76 and 6? 86 and 6? 96 and 6?

10. How many are 6 and 7? 16 and 7? 26 and 7? 36 and 7? 46 and 7? 56 and 7? 66 and 7? 76 and 7? 86 and 7? 96 and 7?

11. How many are 6 and 8? 16 and 8? 26 and 8? 36 and 8? 46 and 8? 56 and 8? 66 and 8? 76 and 8? 86 and 8? 96 and 8?

12. How many are 6 and 9? 16 and 9? 26 and 9? 36 and 9? 46 and 9? 56 and 9? 66 and 9? 76 and 9? 86 and 9? 96 and 9?

13. How many are 6 and 10? 16 and 10? 26 and 10? 36 and 10? 46 and 10? 56 and 10? 66 and 10? 76 and 10? 86 and 10? 96 and 10?

14. How many are 5 and 6? 15 and 6? 25 and 6? 35 and 6? 45 and 6? 55 and 6? 65 and 6? 75 and 6? 85 and 6? 95 and 6?

15. How many are 5 and 7? 15 and 7? 25 and 7? 35 and 7? 45 and 7? 55 and 7? 65 and 7? 75 and 7? 85 and 7? 95 and 7?

16. How many are 5 and 8? 15 and 8? 25 and 8? 35 and 8? 45 and 8? 55 and 8? 65 and 8? 75 and 8? 85 and 8? 95 and 8?

17. How many are 5 and 9? 15 and 9? 25 and 9? 35 and 9? 45 and 9? 55 and 9? 65 and 9? 75 and 9? 85 and 9? 95 and 9?

18 How many are 5 and 10? 15 and 10? 25 and

10? 35 and 10? 45 and 10? 55 and 10? 65 and 10? 75 and 10? 85 and 10? 95 and 10?

19. How many are 4 and 7? 14 and 7? 24 and 7? 34 and 7? 44 and 7? 54 and 7? 64 and 7? 74 and 7? 84 and 7? 94 and 7?

20. How many are 4 and 8? 14 and 8? 24 and 8? 34 and 8? 44 and 8? 54 and 8? 64 and 8? 74 and 8? 84 and 8? 94 and 8?

21. How many are 4 and 9? 14 and 9? 24 and 9? 34 and 9? 44 and 9? 54 and 9? 64 and 9? 74 and 9? 84 and 9? 94 and 9?

22. How many are 4 and 10? 14 and 10? 24 and 10? 34 and 10? 44 and 10? 54 and 10? 64 and 10? 74 and 10? 84 and 10? 94 and 10?

23. How many are 3 and 8? 13 and 8? 23 and 8? 33 and 8? 43 and 8? 53 and 8? 63 and 8? 73 and 8? 83 and 8? 93 and 8?

24. How many are 3 and 9? 13 and 9? 23 and 9? 33 and 9? 43 and 9? 53 and 9? 63 and 9? 73 and 9? 83 and 9? 93 and 9?

25. How many are 3 and 10? 13 and 10? 23 and 10? 33 and 10? 43 and 10? 53 and 10? 63 and 10? 73 and 10? 83 and 10? 93 and 10?

26. How many are 2 and 9? 12 and 9? 22 and 9? 32 and 9? 42 and 9? 52 and 9? 62 and 9? 72 and 9? 82 and 9? 92 and 9?

27. How many are 2 and 10? 12 and 10? 22 and 10? 32 and 10? 42 and 10? 52 and 10? 62 and 10? 72 and 10? 82 and 10? 92 and 10?

LESSON VIII.

1. Thomas paid 33 cents for an arithmetic, 9 cents for a writing-book, and 8 cents for some ink; how many cents did he pay for all?

2. Abel bought a knife for 75 cents, a top for 10 cents, and a ball for 6 cents; what did he pay for all?

3. Thirty-seven apples are growing on one branch, 8 on another, 9 on another, 6 on another, and 9 on another; how many apples are growing on the five branches?

4. Twenty-five peaches were growing on one tree, 6 on another, and 9 on another; but 10 have been picked from the first tree, and 5 from the third; how many peaches remain on the three trees?

5. How many are 26 and 8 and 9 and 7, less 5?

6. A poor woman bought a pound of cheese for 9 cents, a pound of crackers for 8 cents, and 3 pounds of sugar at 11 cents a pound; how many cents did she pay for all?

7. A merchant sold 6 barrels of flour at 8 dollars a barrel, and took his pay in hay at 12 dollars a ton; how many tons did he receive?

8. A farmer has 15 acres of corn, 8 acres of wheat, 5 acres of rye, and 9 acres of oats; how many acres of grain has he?

9. A drover bought 27 cows of one man, 9 of another, 6 of another, and 8 of another. He afterwards sold 10 of them; how many had he then?

10. A sportsman caught 23 squirrels, 3 foxes, 8 rabbits, 10 woodchucks, and 9 birds; how many animals did he catch?

11. In a certain orchard there are 43 apple trees, 10 peach trees, 7 pear trees, 8 cherry trees, and 6 quince trees; how many trees are there in the orchard?

12. A lady bought a silk dress for 22 dollars, a bonnet for 9 dollars, a shawl for 6 dollars, a pair of shoes for 2 dollars, a pair of gloves for 1 dollar, and had 10 dollars left; how many dollars had she at first?

13. A man paid 75 dollars for a horse, 9 dollars for a saddle, 3 dollars for a bridle, and 1 dollar for a whip; what did they all cost him?

14. A miller bought 54 bushels of wheat, 10 of rye,

8 of corn, and 3 of buckwheat; how many bushels of grain did he buy?

15. David has 37 cents in his purse, 8 in one pocket, 9 in another pocket, and seven in his hand; how many cents has he?

16. James had 37 cents, and George gave him 7, Charles 6, Samuel 9, John 8, and he bought a top for 10 cents; how many cents had he left?

17. A lady gave 50 dollars for a watch, 9 dollars for a chain, 1 dollar for a key, and 5 dollars for a pin; how many dollars did she give for all?

18. How many are 54 and 10 and 8 and 6, less 9?

19. How many are 83 and 7 and 9 and 5 and 3, less 8?

LESSON IX.

1. How many are 11 less 2? 21 less 2? 31 less 2? 41 less 2? 51 less 2? 61 less 2? 71 less 2? 81 less 2? 91 less 2? 101 less 2?

2. How many are 11 less 3? 21 less 3? 31 less 3? 41 less 3? 51 less 3? 61 less 3? 71 less 3? 81 less 3? 91 less 3? 101 less 3?

3. How many are 11 less 9? 21 less 9? 31 less 9? 41 less 9? 51 less 9? 61 less 9? 71 less 9? 81 less 9? 91 less 9? 101 less 9?

4. How many are 12 less 5? 22 less 5? 32 less 5? 42 less 5? 52 less 5? 62 less 5? 72 less 5? 82 less 5? 92 less 5? 102 less 5?

5. How many are 14 less 8? 24 less 8? 34 less 8? 44 less 8? 54 less 8? 64 less 8? 74 less 8? 84 less 8? 94 less 8? 104 less 8?

6. Thirty-six and 9 and 4 and 8, less 7, are how many?

7. Forty-four and 2 and 7 and 9 and 3 and 8 and 5, less 6, are how many?

8. Fifty-nine and 5 and 6 and 7 and 3 and 8 and 4, less 2, are how many?

9. Sixteen and 4 and 10 and 12 and 8 and 3 and 9, less 7, are how many?

10. Eighty-one and 9 and 10 and 3 and 8 and 6, less 7, are how many?

LESSON X.

1. WHAT cost 9 yards of cloth, at $4 a yard?

2. What cost 8 pounds of sugar, at 9 cents a pound?

3. What cost 6 dozen of eggs, at 12 cents a dozen?

4. A and B start from the same place and travel in the same direction, A at the rate of 9 miles an hour and B 6 miles; how far apart are they in 1 hour? In 5 hours? In 8 hours? 10 hours? 12 hours? 20 hours?

5. A and B start from the same place and travel in opposite directions, A at the rate of 5 miles an hour, and B 4 miles; how far apart are they in 1 hour? In 6 hours? 9 hours? 10 hours? 12 hours?

6. A man bought 9 yards of cloth, at 5 dollars a yard, for a suit of clothes, and paid 6 dollars for making the coat, 2 dollars for making the pantaloons, and 1 dollar for making the vest; what did the suit cost?

7. At 9 cents each, what will 6 pineapples cost?

8. At 5 dollars a pair, what will 9 pair of boots cost? 6 pair? 10 pair? 12 pair?

9. In my garden there are 7 rows of corn, and 9 hills in each row; how many hills of corn are there in the garden?

10. In 1 bushel there are 4 pecks; how many pecks are there in 3 bushels? In 5 bushels?

11. How many pecks are there in 5 bushels and 1 peck? In 5 bushels and 3 pecks?

12. In 1 peck there are 8 quarts; how many quarts are there in 4 pecks? Then how many quarts are there in 1 bushel? Why?

13. How many quarts are there in 1 bushel and 1 peck? How many in 1 bushel and 2 pecks? How many in 1 bushel, 3 pecks, and 2 quarts?

14. Count by twos from 50 to 100; thus, 50, 52, 54, etc.

15. Count backward from 100 thus: 100, 98, 96, etc.

16. Count by twos from 51 to 101; thus, 51, 53, 55, etc.

17. Count backward from 101 thus: 101, 99, 97, etc.

18. Count by threes to 99; thus, 3, 6, 9, etc.

19. Count backward from 99 thus: 99, 96, 93, etc.

20. Count by threes from 1 to 100; thus, 1, 4, 7, 10, etc.

21. Count backward from 100 thus: 100, 97, 94, etc.

22. Count by threes from 2 to 101; thus, 2, 5, 8, etc.

23. Count backward from 101 thus: 101, 98, 95, etc.

LESSON XI.

1. IN one pint there are 4 gills; how many gills are there in 3 pints? In 5 pints? In 2 pints and 3 gills?

2. In one quart there are 2 pints; how many pints are there in 3 quarts? In 5 quarts? In 7 quarts and 1 pint?

3. How many gills are there in 1 quart? In 3 quarts? In 5 quarts and 1 pint? In 6 quarts, 1 pint, and 3 gills?

4. If wine costs 5 cents a gill, what will one quart cost? What will one quart and one pint cost? One quart, one pint, and two gills?

5. In one gallon there are 4 quarts; how many quarts are there in 3 gallons? In 5 gallons? In 1 gallon and 3 quarts?

6. How many pints are there in 1 gallon? In 3 gallons? In 1 gallon and 3 quarts?

7. How many pints are there in 1 gallon, 3 quarts, and 1 pint?

8. When molasses is worth 7 cents a pint, what is 1 gallon worth? What is 1 gallon and 1 quart worth? 1 gallon, 2 quarts, and 1 pint?

9. How many gills are there in 1 gallon?

10. How many gills are there in 2 gallons and 2 quarts?

11. How many gills are there in 1 gallon, 2 quarts, and 1 pint? In 2 gallons, 1 quart, 1 pint, and 3 gills?

12. At 2 cents a pint, what will 3 quarts of milk cost? 3 quarts and 1 pint?

13. At 3 cents a gill, what will 1 pint of wine cost? 1 quart? 2 quarts and 1 pint? 3 quarts, 1 pint, and 2 gills?

14. At 4 cents a pint, what will 1 gallon of gin cost?

15. At 10 cents a quart, what will 1 gallon of molasses cost? 2 gallons and 3 quarts?

16. At 10 cents a gill, what will 3 pints of brandy cost? 1 quart, 1 pint, and 3 gills?

17. If one peck of wheat costs 3 shillings, how many shillings will one bushel cost? How many dollars, if 6 shillings make 1 dollar?

18. If cherries cost 8 cents a quart, what will 1 peck cost? What 1 peck and 2 quarts?

19. When beans are worth 5 cents a pint, what is 1 peck worth? What 6 quarts and 1 pint?

LESSON XII.

1. How many are 7 times 1?	Once 7?
2. How many are 7 times 2?	Twice 7?
3. How many are 7 times 3?	3 times 7?
4. How many are 7 times 4?	4 times 7?
5. How many are 7 times 5?	5 times 7?
6. How many are 7 times 6?	6 times 7?
7. How many are 7 times 7?	

8. How many are 7 times 8? 8 times 7?
9. How many are 7 times 9? 9 times 7?
10. How many are 7 times 10? 10 times 7?
11. How many are 7 times 11? 11 times 7?
12. How many are 7 times 12? 12 times 7?
13. How many are 8 times 1? Once 8?
14. How many are 8 times 2? Twice 8?
15. How many are 8 times 3? 3 times 8?
16. How many are 8 times 4? 4 times 8?
17. How many are 8 times 5? 5 times 8?
18. How many are 8 times 6? 6 times 8?
19. How many are 8 times 7? 7 times 8?
20. How many are 8 times 8?
21. How many are 8 times 9? 9 times 8?
22. How many are 8 times 10? 10 times 8?
23. How many are 8 times 11? 11 times 8?
24. How many are 8 times 12? 12 times 8?
25. How many are 9 times 1? Once 9?
26. How many are 9 times 2? Twice 9?
27. How many are 9 times 3? 3 times 9?
28. How many are 9 times 4? 4 times 9?
29. How many are 9 times 5? 5 times 9?
30. How many are 9 times 6? 6 times 9?
31. How many are 9 times 7? 7 times 9?
32. How many are 9 times 8? 8 times 9?
33. How many are 9 times 9?
34. How many are 9 times 10? 10 times 9?
35. How many are 9 times 11? 11 times 9?
36. How many are 9 times 12? 12 times 9?
37. Count by 4's to 100; thus, 4, 8, 12, etc.
38. Count backward by 4's; thus, 100, 96, 92, etc.
39. Count by 5's to 100; thus, 5, 10, 15, etc.
40. Count backward by 5's; thus, 100, 95, 90, etc.
41. Take 7, add 5, multiply by 4, add 2, divide by 5, subtract 2, multiply by 4, add 4, divide by 9, divide by 2, multiply by 8, subtract 1, divide by 3, multiply by 4, add 5; result?

LESSON XIII.

1. If two men can cut a certain field of wheat in four days, how many days will it take one man to cut it?

2. If three men can do a piece of work in five days, in how many days could one man do the same?

3. If a quantity of provisions will serve five men six days, how many days would the same provisions serve one man?

4. If a quantity of provisions will serve four men five days, how many men would the same provisions serve one day?

5. Six men built a piece of fence in seven days, but, the fence being destroyed, it is necessary to rebuild it in one day; how many men must be employed to do the work?

6. Paid twenty dollars for five yards of cloth; what was the price of one yard?

7. If a horse can travel 6 miles in one hour, in how many hours can he travel 18 miles?

8. A butcher bought sheep at 7 dollars a head; how many did he buy for 28 dollars? For 42 dollars?

9. At 8 dollars a barrel, how many barrels of flour can I buy for 24 dollars? For 40 dollars?

10. At 7 dollars a barrel, what are 5 barrels of flour worth?

11. A laborer worked 5 months for 60 dollars; what did he receive for 1 month? For 3 months?

12. If the interest of 1 dollar is 6 cents for a year, what is the interest of 7 dollars for the same time?

13. If the interest of 1 dollar is 6 cents for 1 year, what is the interest of the same sum for 9 years? For 12 years?

14. If a man can earn six shillings in a day, how many shillings can he earn in 5 days? In 8 days?

15. At $5 a week, what will 8 weeks' board cost?

16. A man can earn 10 dollars in a month; how much can he earn in 7 months? In 9 months?

17. A man worked for 11 dollars a month, and earned 77 dollars; how long did he work?

18. If 72 peaches are divided equally between 8 boys, how many does each receive?

19. How many acres of land, at 12 dollars an acre, may be bought for 36 dollars? For 60 dollars?

20. Seven days make a week; how many days in 9 weeks? In 12 weeks?

21. A man can perform a journey in 63 hours; in how many days, of 9 hours each, can he perform it?

LESSON XIV.

1. EIGHT are how many times 2? 4?
2. Ten are how many times 5? 2? 10? 1?
3. Twelve are how many times 2? 3? 6? 4?
4. Fourteen are how many times 2? 7? 14?
5. Fifteen are how many times 3? 5? 1?
6. Sixteen are how many times 2? 4? 8?
7. Eighteen are how many times 6? 9? 2? 3?
8. Twenty are how many times 4? 10? 2? 5?
9. Twenty-one are how many times 3? 7?
10. Twenty-four are how many times 3? 6? 8? 2?
11. Twenty-eight are how many times 4? 2? 7? 14?
12. Thirty are how many times 5? 10? 6? 3? 2?
13. Thirty-two are how many times 8? 4? 2? 16?
14. Forty are how many times 8? 4? 5? 10?
15. Forty-two are how many times 6? 7? 3? 21?
16. Forty-five are how many times 3? 9? 5? 15?
17. Forty-eight are how many times 4? 8? 12? 6?
18. Fifty are how many times 5? 25? 10? 2?
19. Fifty-four are how many times 6? 3? 9? 2?
20. Sixty-four are how many times 8? 4? 16?
21. Seventy-two are how many times 8? 6? 9?

22. If 4 horses eat 8 tons of hay in 6 months, how many horses will eat 8 tons in 12 months?

23. If 5 horses eat 10 tons of hay in 6 months, how many tons will 7 horses eat in the same time?

24. Bought 6 tons of coal at 8 dollars a ton, and paid for it with cloth at 4 dollars a yard; how many yards did it take?

25. If 6 men can do a piece of work in 10 days, how many men can do the same in 15 days? In 5 days?

26. If 6 men can do a piece of work in 15 days, in how many days can 10 men do the same? 30 men?

27. How many barrels of pork, at 12 dollars a barrel, will pay for 10 barrels of flour, at 6 dollars a barrel?

28. How many 8's in 16? In 48? In 72?
29. How many 7's in 35? 63? 49? 28?
30. How many 9's in 36? 54? 27? 81?
31. How many 6's in 48? 60? 72? 36?
32. How many 10's in 30? 70? 90? 110?
33. How many 5's in 50? 35? 60? 25?

LESSON XV.

READ across the page thus: 5 and 1 are 6; 5 from 6 leaves 1; 5 times 1 are 5; 5 in 5, once; etc.

5 and 1?	5 from 6?	5 times 1?	5 in 5?
5 and 2?	5 from 7?	5 times 2?	5 in 10?
5 and 3?	5 from 8?	5 times 3?	5 in 15?
5 and 4?	5 from 9?	5 times 4?	5 in 20?
5 and 5?	5 from 10?	5 times 5?	5 in 25?
5 and 6?	5 from 11?	5 times 6?	5 in 30?
5 and 7?	5 from 12?	5 times 7?	5 in 35?
5 and 8?	5 from 13?	5 times 8?	5 in 40?
5 and 9?	5 from 14?	5 times 9?	5 in 45?
5 and 10?	5 from 15?	5 times 10?	5 in 50?
5 and 11?	5 from 16?	5 times 11?	5 in 55?
5 and 12?	5 from 17?	5 times 12?	5 in 60?

6 and 1?	6 from 7?	6 times 1?	6 in 6?
6 and 2?	6 from 8?	6 times 2?	6 in 12?
6 and 3?	6 from 9?	6 times 3?	6 in 18?
6 and 4?	6 from 10?	6 times 4?	6 in 24?
6 and 5?	6 from 11?	6 times 5?	6 in 30?
6 and 6?	6 from 12?	6 times 6?	6 in 36?
6 and 7?	6 from 13?	6 times 7?	6 in 42?
6 and 8?	6 from 14?	6 times 8?	6 in 48?
6 and 9?	6 from 15?	6 times 9?	6 in 54?
6 and 10?	6 from 16?	6 times 10?	6 in 60?
6 and 11?	6 from 17?	6 times 11?	6 in 66?
6 and 12?	6 from 18?	6 times 12?	6 in 72?
7 and 1?	7 from 8?	7 times 1?	7 in 7?
7 and 2?	7 from 9?	7 times 2?	7 in 14?
7 and 3?	7 from 10?	7 times 3?	7 in 21?
7 and 4?	7 from 11?	7 times 4?	7 in 28?
7 and 5?	7 from 12?	7 times 5?	7 in 35?
7 and 6?	7 from 13?	7 times 6?	7 in 42?
7 and 7?	7 from 14?	7 times 7?	7 in 49?
7 and 8?	7 from 15?	7 times 8?	7 in 56?
7 and 9?	7 from 16?	7 times 9?	7 in 63?
7 and 10?	7 from 17?	7 times 10?	7 in 70?
7 and 11?	7 from 18?	7 times 11?	7 in 77?
7 and 12?	7 from 19?	7 times 12?	7 in 84?
8 and 1?	8 from 9?	8 times 1?	8 in 8?
8 and 2?	8 from 10?	8 times 2?	8 in 16?
8 and 3?	8 from 11?	8 times 3?	8 in 24?
8 and 4?	8 from 12?	8 times 4?	8 in 32?
8 and 5?	8 from 13?	8 times 5?	8 in 40?
8 and 6?	8 from 14?	8 times 6?	8 in 48?
8 and 7?	8 from 15?	8 times 7?	8 in 56?
8 and 8?	8 from 16?	8 times 8?	8 in 64?
8 and 9?	8 from 17?	8 times 9?	8 in 72?
8 and 10?	8 from 18?	8 times 10?	8 in 80?
8 and 11?	8 from 19?	8 times 11?	8 in 88?
8 and 12?	8 from 20?	8 times 12?	8 in 96?

SECTION THIRD.

LESSON I.

REMARK. When any thing is divided into *two equal parts*, one of the parts is called *one half* of the thing; so when any *number* is divided into *two equal parts*, one of these parts is called *one half* of the number.

1. If an apple is worth two cents, what is one half of it worth?

Ans. It is worth *one half* of two cents.

2. What is one half of two cents?

Ans. One half of two cents is one cent.

3. Why? *Ans.* Because, if we divide two cents into two equal parts, one of the parts is one cent.

4. How many halves of an apple make a whole apple?

5. How many halves of a *pear* make a whole *pear*?

6. Are two halves of a *dollar* equal to a *whole dollar*?

Ans. Yes. Two halves of *any thing* make the whole of *that thing*.

7. If you can buy a pear for two cents, what part of it can you buy for one cent?

8. One is what part of two?

Ans. One is *one half* of two.

9. If you can buy one pear for two cents, how many pears can you buy for three cents?

10. Three are how many times two?

Ans. Once two and one half of two.

11. If you can buy one yard of cloth for two dollars, how many yards can you buy for four dollars?

12. Four are how many times two?

13. Two is what part of four?

14. Six are how many times two?

15. Five are how many times two?

Ans. Two times two and one half of two.

16. If two dollars will buy one barrel of apples, how many barrels will five dollars buy?

17. If two dollars will buy a bushel of wheat, how many bushels will seven dollars buy?

18. Eight are how many times two?

19. Nine are how many times two?

20. Ten are how many times two?

REMARK. When any thing or any number is divided into *three equal parts*, one of these parts is called *one third* of the thing or *one third* of the number; *two* of the parts are *two thirds* of the thing or of the number.

21. How many thirds of an apple make a whole apple?

22. If a load of wood is worth three dollars, and if it is divided into three equal parts, what will one of the parts be worth; that is, what will one third of the load be worth?

23. What is one third of three dollars?

24. Suppose the load to be divided as before, what will two of the parts be worth; that is, what will two thirds of the load be worth?

25. What is two thirds of three?

26. If a barrel of apples is worth three dollars, what part of a barrel will one dollar buy? What part of a barrel will two dollars buy?

27. One is what part of three?

Ans. One is *one third* of three.

28. Two is what part of three?

Ans. Two is two times one third of three; that is, two is *two thirds* of three.

29. If you can buy a pound of tea for three shillings, how much can you buy for four shillings? How much for five shillings? For six shillings?

30. Four are how many times three?

Ans. Once three and one third of three.

31. Five are how many times three?

32. Six are how many times three?

33. If apples are worth three dollars a barrel, how many barrels can you buy for seven dollars? How many for eight dollars?

34. What is meant by one third of any thing?

Ans. One third of any thing is one of the three equal parts into which that thing is divided. See Remark after Example 20.

35. What is meant by two thirds of any thing?

36. Nine are how many times three?

37. Ten are how many times three?

LESSON II.

1. WHAT is meant by one fourth of a thing?

Ans. One of the four equal parts into which the thing is divided.

2. What is meant by two fourths, or three fourths, of any thing or of any number?

3. If a cake is worth 4 cents, and it is cut into 4 equal parts, what is 1 of the parts worth; that is, what is 1 fourth of the cake worth? What are 2 fourths of it worth? What are 3 fourths of it worth?

4. If you can buy a cord of wood for 4 dollars, what part of a cord can you buy for 1 dollar? What part for 2 dollars? For 3 dollars?

5. What part of four is one?

Ans. One is *one fourth* of four.

6. What part of four is two?

Ans. *Two fourths*, or *one half*, of four.

7. What part of four is three?

Ans. *Three fourths* of four.

8. How many fourths make a whole one?

9. If you can buy a yard of cloth for four dollars, how much can you buy for 5 dollars? How much for 6 dollars? For 7 dollars?

10. Five are how many times four?

Ans. Once 4 and one fourth of 4.

11. Six are how many times 4?

Ans. Once 4 and two fourths of 4; or, once 4 and one half of 4.

12. Seven are how many times 4?

Ans. Once 4 and three fourths of 4.

13. Eight are how many times 4?

14. If 4 barrels of apples will buy 1 ton of coal, how many tons will 9 barrels buy? How many tons will 10 barrels buy? How many tons will 11 barrels buy?

15. Twelve are how many times 4?

16. Thirteen are how many times 4?

17. Fourteen are how many times 4?

18. Fifteen are how many times 4?

19. Sixteen are how many times 4?

20. What is one fourth of 4?

Ans. One fourth of 4 is 1.

21. Why? *Ans.* Because, if we divide 4 into *four* equal parts, *one* of the parts is 1.

22. What is two fourths, or one half, of 4.

23. What is three fourths of 4?

24. If one half of an apple costs 1 cent, what will the whole apple cost?

Solution. Since *one* half of the apple costs 1 cent, *two* halves, or the whole apple, will cost *two times* 1 cent, which are 2 cents.

25. If one half of a barrel of flour is worth 4 dollars, what is a whole barrel worth? Why?

26. If one third of a melon is worth 4 cents, what are two thirds of it worth? What is the whole melon worth?

27. If one fourth of a pie costs 5 cents, what will two fourths, or one half, of the pie cost? What will three fourths of it cost? What the whole pie?

28. What is one half of 6 apples?

Ans. Three apples.

29. Why? *Ans.* Because, if 6 apples are divided into two equal parts, each part is 3 apples.

50. What is one half of 8 cents? Why?

31. What is one half of 12 oranges? Why?

32. What is one half of 10? Why?

33. What is one third of 6 apples?

Ans. Two apples.

34. Why? *Ans.* Because, if 6 apples are divided into three equal parts, each part is 2 apples.

35. What is one third of 12 lemons? Why?

36. What is two thirds of 12 lemons? Why?

37. What is one third of 15 dollars? What is two thirds of 15 dollars?

38. What is one third of 21? Two thirds of 21?

39. What is one half of 3 apples?

Ans. One apple and one half of another apple.

40. What is one half of 5 apples?

Ans. Two apples and one half of an apple.

41. What is one half of 7 apples?

42. What is one half of 9 pears?

43. What is one third of 4 oranges? Two thirds?

44. What is one fourth of 16 miles?

45. What is two fourths, or one half, of 16 miles?

46. What is three fourths of 16 miles?

47. What is one fifth of 15 dollars?

48. What is two fifths of 15 dollars? Three fifths?

49. What is meant by one fifth, two fifths, etc., of any number? See Remark after Ex. 20, page 49.

50. Mary had two fifths of a dollar, and her mother gave her another fifth; what part of a dollar had she then?

51. Henry had four fifths of a dollar, but he spent two fifths of a dollar for fruit; what part of a dollar has he now?

52. How many fifths make a whole one?

53. A boy spent one fifth of all his money on Monday, and two fifths on Tuesday; what part of his money did he spend in these two days? What part had he then remaining?

LESSON III.

REMARK. The following lines represent a yard-stick, marked off or divided into *halves*, *fourths*, and *eighths* of a yard.

One half.				One half.			
One fourth.		One fourth.		One fourth.		One fourth.	
1 eighth.	1 eighth.	1 eighth.	1 eighth.	1 eighth.	1 eighth.	1 eighth.	1 eighth

1. One half equals how many fourths? *Ans.* Two.
2. One fourth equals how many eighths?
3. One half equals how many eighths?
4. If you cut an apple into *halves*, and then cut *each half* into two equal pieces, into how many pieces will the apple be cut? What part of the whole apple is one of these *small pieces*?
5. One half of an apple is equal to how many times one fourth of an apple?
6. How many fourths of an orange are equal to one half of an orange?

REMARK. The following lines represent a yard-stick divided into *thirds* and *sixths* of a yard.

One third.		One third.		One third.	
One sixth.	One sixth.	One sixth.	One sixth.	One sixth.	One sixth.

7 One third equals how many sixths?
8. If you cut a cake into *thirds*, and then cut *each third* into two equal parts, into how many equal parts will the cake be cut? Each of the small pieces will be what part of the whole cake?
9. One third of a cake is how many times one sixth of the same cake?
10. If you cut a pie into *halves*, and then cut *each half* into three equal parts, what part of the pie will one of these small pieces be?

11. One half of any thing is how many sixths of the same thing?

12. How many sixths of an apple are equal to one third of an apple?

13. How many sixths of an apple are equal to one half of an apple?

14. If *one half* of an apple is cut into *four* equal pieces, what part of the whole apple will one of these small pieces be?

15. How many eighths of an apple make one half of an apple?

16. If *one fourth* of an apple is cut into *two* equal parts, what part of the whole apple will one of these small pieces be?

17. How many eighths of a pear make one fourth of a pear?

18. What is meant by one eighth of any thing?

19. If a barrel of flour be worth 8 dollars, and if it be divided equally among 8 men, what will one man's share be worth; that is, what is one eighth of a barrel worth? What are two eighths, or one fourth, of it worth? What are three eighths of it worth? What are four eighths, or one half, of it worth?

20. If 8 dollars will buy a box of tea, what part of a box will 1 dollar buy? What part will 2 dollars buy? 3 dollars? 4 dollars? 5 dollars? 6 dollars? 7 dollars?

21. What part of eight is one?

Ans. One is one eighth of eight.

22. Two is what part of eight?

Ans. Two eighths, or one fourth of eight.

23. Three is what part of eight?

Ans. Three eighths of eight.

24. Four is what part of eight?

Ans. Four eighths, or one half of eight.

25. Five is what part of eight?

26. Six is what part of eight?

27. Seven is what part of eight?

28. Nine are how many times eight?

Ans. Once eight and one eighth of eight.

29. Ten are how many times eight?

Ans. Once eight and two eighths of eight, or once eight and one fourth of eight.

NOTE. The second answer to Example 29 is the best, if the pupil is able to make the reduction. The *teacher* should exercise his own judgment in all such questions.

30. Eleven are how many times eight?
31. Twelve are how many times eight?
32. Thirteen are how many times eight?
33. Fourteen are how many times eight?
34. Fifteen are how many times eight?

LESSON IV.

1. IF cranberries cost 7 cents a quart, what part of a quart may be bought for 1 cent? What part for 2 cents? For 3 cents? 4 cents? 5 cents? 6 cents?

2. If wood is worth 7 dollars a cord, how many cords may be bought for 8 dollars? For 9 dollars? 10 dollars? 12 dollars? 14 dollars?

3. What is meant by one seventh, two sevenths, three sevenths, etc., of any thing?

4. How many sevenths make a whole one?

5. If corn is worth 7 shillings a bushel, what is one seventh of a bushel worth? What are two sevenths worth? Three sevenths? Five sevenths?

6. If blueberries cost 7 cents a quart, what will one seventh of a quart cost? Four sevenths? Six sevenths?

7. One is what part of seven?

Ans. One is one seventh of seven.

8. Two is what part of seven?

Ans. Two sevenths of seven.

9. Three is what part of seven?

10. Nine are how many times seven?

Ans. Once seven and two sevenths of seven.

11. Ten are how many times seven?
12. Twelve are how many times seven?
13. Fourteen are how many times seven?
14. Fifteen are how many times seven?
15. When rye is worth 9 shillings a bushel, what is one ninth of a bushel worth? What are two ninths of a bushel worth? What are three ninths, or one third of a bushel worth? Four ninths? Five ninths? Six ninths, or two thirds? Seven ninths? Eight ninths?
16. When coal costs 7 dollars a ton, what part of a ton can I buy for 1 dollar? What part for 2 dollars? For 3 dollars? For 4 dollars? For 5 dollars? For 6 dollars? For 7 dollars? How much can I buy for 8 dollars? For 10 dollars? For 11 dollars? For 12 dollars? For 14 dollars? For 16 dollars?
17. One is what part of nine?
18. What is meant by one ninth, two ninths, three ninths, etc.?
19. How many ninths make a whole one?
20. Two is what part of nine?
21. Three is what part of nine?
22. Four is what part of nine?
23. Five is what part of nine?
24. Six is what part of nine?
25. Seven is what part of nine?
26. Eight is what part of nine?
27. Ten are how many times nine?
28. Eleven are how many times nine?
29. Twelve are how many times nine?
30. Fourteen are how many times nine?
31. Fifteen are how many times nine?
32. Seventeen are how many times nine?
33. Eighteen are how many times nine?
34. Twenty are how many times nine?
35. Twenty-four are how many times nine?

LESSON V.

1. If a laborer earns 3 shillings in a half day, how much will he earn in a whole day?

2. Three is one half of what number?

Solution. If 3 is *one* half of some number, then *two* halves, or the whole of the number, must be twice 3, which are 6; therefore 3 is one half of 6?

3. Five is one half of what number? Why?

4. If a boy picks 2 quarts of chestnuts in 1 hour, how many quarts will he pick in 3 hours?

5. Two is one third of what number?

Solution. If 2 is *one* third of some number, then *three* thirds, or the whole of the number, must be three times 2, which are 6; therefore, 2 is one third of 6.

6. Four is one third of what number? Why?

7. Seven is one third of what number?

8. If one fourth of a dollar will pay for 6 oranges, how many oranges may be bought for a whole dollar?

9. Six is one fourth of what number?

10. Nine is one fourth of what number?

11. Three is one fifth of what number?

12. Five is one seventh of what number?

13. Four is one eighth of what number?

14. If three fourths of a pound of raisins cost 15 cents, what will a pound cost?

Solution. If *three* fourths cost 15 cents, *one* fourth will cost one third of 15 cents, which is 5 cents; and *four* fourths will cost 4 times 5 cents, which are 20 cents.

15. Charlie's mother said to him, "I have just bought 5 eighths of a watermelon for 15 cents; and if you will tell me what the whole melon is worth, I will give you money enough to buy one." Had you been Charlie, what would you have answered?

16. Fifteen is 3 fourths of what number?

17. Fifteen is 5 eighths of what number?

18. Twelve is 3 fifths of what number?

19. Twenty is 5 sevenths of what number?

20. Eighteen is 6 sevenths of what number?

21. When cherries cost 10 cents a quart, what is one tenth of a quart worth? What are two tenths, or one fifth, of a quart worth? What are three tenths worth? What four tenths, or two fifths? What five tenths, or one half?

22. When hay costs 10 dollars a ton, what part of a ton can you buy for 1 dollar? What part for 2 dollars? What for 3 dollars? For 4 dollars? 5 dollars? 6 dollars? 7 dollars? 8 dollars? How much can you buy for 11 dollars? For 12 dollars? 13 dollars? 15 dollars? 19 dollars? 20 dollars?

23. What is meant by one tenth, two tenths, three tenths, etc.?

24. How many tenths make a whole one?

25. If a barrel of flour costs 10 dollars, what will one half of a barrel cost? What will 1 fifth of a barrel cost? 3 fifths? 4 fifths?

26. If I can walk 4 miles in 1 hour, how far can I walk in a half hour? How far in 3 fourths of an hour?

27. If a man earns 8 shillings each day, in how many days will he earn 25 shillings? 34 shillings? 36 shillings?

28. If a man threshes 10 bushels of rye each day, in how many days will he thresh 35 bushels? 44 bushels? 28 bushels?

29. When hay is worth 12 dollars a ton, how many tons can I buy for 30 dollars? For 33 dollars?

LESSON VI.

1. TEN are how many times 2? How many times 5? 4? 3?

2. Fifteen are how many times 3? 5? 6?

3. Sixteen are how many times 4? 8? 6? 5?

4. Eighteen are how many times 6? 9? 3? 2? 4?

5. Twenty are how many times 5? 10? 4? 2? 8?

6. Nineteen are how many times 6? 3? 11? 10? 2? 5?

7. Twelve are how many times 4? 6? 3? 2? 5? 8?

8. Seventeen are how many times 3? 5? 4? 2? 6? 8?

9. Thirteen are how many times 4? 3? 5? 2? 6? 9? 7?

10. Twenty-one are how many times 7? 3? 6? 8? 2? 5? 4?

11. If you had 22 cents, how many oranges could you buy, at 4 cents apiece? How many at 3 cents? At 5 cents? At 2 cents?

12. Twenty-two are how many times 4? 3? 5? 2? 11? 7? 6?

13. Twenty-five are how many times 5? 2? 10? 6? 4?

14. Twenty-three are how many times 7? 5? 4? 10? 2?

15. Twenty-four are how many times 6? 8? 4? 3? 12? 5?

16. Twenty-eight are how many times 4? 8? 7? 6? 9?

17. Twenty-six are how many times 2? 13? 6? 4? 8?

18. Thirty are how many times 6? 10? 3? 5? 8? 4? 7?

19. Twenty-seven are how many times 9? 6? 3? 5? 7?

20. Twenty-nine are how many times 7? 3? 5? 6? 8?

21. Thirty-two are how many times 8? 6? 4? 5? 2? 16?

22. Thirty-one are how many times 10? 6? 4? 3? 8? 7?

LESSON VII.

1. IF you had $32, how many yards of cloth could you, buy at $4 a yard? At $2? $6? $5? $8? $1? $3? $7?

2. Thirty-three are how many times 10? 11? 6? 4? 3? 12?

3. Thirty-four are how many times 6? 10? 5? 8? 3? 4?

4. Thirty-five are how many times 7? 10? 8? 5? 4? 6? 3?

5. For 36 cents, how many pounds of sugar can be bought, at 6 cents a pound? At 12 cents? 10 cents? 8 cents? 9 cents? 7 cents? 11 cents?

6. Thirty-six are how many times 6? 12? 10? 8? 9? 7? 11?

7. Forty are how many times 8? 6? 4? 10? 9? 7? 12? 11?

8. Forty-five are how many times 9? 8? 6? 5? 10? 7? 3?

9. Forty-eight are how many times 12? 6? 4? 8? 9?

10. Fifty are how many times 10? 8? 5? 4? 12? 6? 7? 9?

11. Thirty-seven are how many times 6? 8? 10? 12? 4? 9?

12. Forty-six are how many times 7? 8? 4? 10? 9? 5? 11?

13. Forty-two are how many times 7? 10? 4? 6? 8? 5? 9?

14. Forty-nine are how many times 7? 8? 6? 10? 9? 4? 12?

15. Thirty-eight are how many times 12? 6? 4? 9? 7? 8?

16. Forty-four are how many times 11? 8? 4? 9? 6? 10?

17. Thirty-nine are how many times 6? 9? 12?

18. For 47 cents, how many oranges may be bought, at 5 cents each? At 4 cents? 8 cents? 6 cents? 9 cents? 7 cents?

19. Forty-three are how many times 6? 8? 10? 5? 9? 7?

20. Forty-one are how many times 10? 12? 4? 6? 9? 7?

LESSON VIII.

1.	How many are 10 times 1?	Once 10?
2.	How many are 10 times 2?	Twice 10?
3.	How many are 10 times 3?	3 times 10?
4.	How many are 10 times 4?	4 times 10?
5.	How many are 10 times 5?	5 times 10?
6.	How many are 10 times 6?	6 times 10?
7.	How many are 10 times 7?	7 times 10?
8.	How many are 10 times 8?	8 times 10?
9.	How many are 10 times 9?	9 times 10?
10.	How many are 10 times 10?	
11.	How many are 10 times 11?	11 times 10?
12.	How many are 10 times 12?	12 times 10?
13.	How many are 11 times 1?	Once 11?
14.	How many are 11 times 2?	Twice 11?
15.	How many are 11 times 3?	3 times 11?
16.	How many are 11 times 4?	4 times 11?
17.	How many are 11 times 5?	5 times 11?
18.	How many are 11 times 6?	6 times 11?
19.	How many are 11 times 7?	7 times 11?
20.	How many are 11 times 8?	8 times 11?
21.	How many are 11 times 9?	9 times 11?
22.	How many are 11 times 10?	10 times 11?
23.	How many are 11 times 11?	
24.	How many are 11 times 12?	12 times 11?
25.	How many are 12 times 1?	Once 12?
26.	How many are 12 times 2?	Twice 12?

27. How many are 12 times 3? 3 times 12?
28. How many are 12 times 4? 4 times 12?
29. How many are 12 times 5? 5 times 12?
30. How many are 12 times 6? 6 times 12?
31. How many are 12 times 7? 7 times 12?
32. How many are 12 times 8? 8 times 12?
33. How many are 12 times 9? 9 times 12?
34. How many are 12 times 10? 10 times 12?
35. How many are 12 times 11? 11 times 12?
36. How many are 12 times 12?

LESSON IX.

1. Two times two are how many times one?
2. Three times two are how many times one?
3. Five times two are how many times one?
4. Eight times two are how many times one?
5. Two times three are how many times one?
6. Five times three are how many times one?
7. Four times three are how many?
8. Seven times two are how many?
9. Nine times three are how many?
10. Ten times two are how many?
11. Six times four are how many?
12. Seven times three are how many?
13. Ten times four are how many?
14. Eleven times two are how many?
15. Five times five are how many?
16. Six times eight are how many?
17. Eight times six are how many?
18. Five times nine are how many?
19. Four times eleven are how many?
20. Four times twelve are how many?
21. Six times twelve are how many?
22. Six times seven are how many?
23. Nine times four are how many?
24. Ten times nine are how many?

25. Eleven times six are how many?
26. Eight times eight are how many?
27. Seven times twelve are how many?
28. Twelve times nine are how many?
29. Twelve times six are how many?
30. Four times nine are how many?
31. Seven times eleven are how many?
32. Nine times eight are how many?
33. Ten times twelve are how many?
34. Eight times three are how many?
35. Six times nine are how many?
36. Twelve times eight are how many?
37. Seven times nine are how many?
38. Seven times ten are how many?
39. Eleven times eight are how many?
40. Seven times eight are how many?
41. Nine times nine are how many?
42. Eleven times nine are how many?
43. Eight times twelve are how many?
44. Five times twelve are how many?
45. Six times eleven are how many?
46. Twelve times three are how many?
47. Nine times five are how many?
48. Eight times seven are how many?
49. Twelve times seven are how many?
50. Five times eight are how many?
51. Nine times ten are how many?
52. Nine times six are how many?
53. Eleven times seven are how many?
54. Three times nine are how many?
55. Eight times eleven are how many?
56. Twelve times twelve are how many?
57. Eight times four are how many?
58. Ten times six are how many?
59. Three times eight are how many?
60. Six times ten are how many?
61. Eleven times ten are how many?

62. Nine times twelve are how many?
63. Five times eleven are how many?
64. Twelve times ten are how many?
65. Nine times eleven are how many?
66. Nine times seven are how many?
67. Six times six are how many?
68. Twelve times four are how many?
69. Ten times eleven are how many?
70. Eleven times twelve are how many?
71. Seven times six are how many?
72. Eleven times eleven are how many?
73. Eight times ten are how many?
74. Seven times seven are how many?
75. Twelve times five are how many?
76. Ten times three are how many?
77. Ten times seven are how many?
78. Eight times nine are how many?
79. Ten times ten are how many?
80. Twelve times eleven are how many?

LESSON X.

1. WHAT cost 5 oranges, at 6 cents apiece?
2. What cost 8 tons of coal, at 7 dollars a ton?
3. There is an orchard consisting of 8 rows of trees, having 12 trees in each row; how many trees are there in the orchard?
4. Twenty shillings make one pound; how many pence are there in a pound?
5. How many gills are there in two gallons?
6. If one gill of beer costs two cents, what will one gallon cost?
7. Two men start from the same place and travel the same way; one travels at the rate of three miles in an hour, and the other seven; how far apart are they at the end of one hour? How far at the end of seven hours?

8. Two men start from the same place and travel in opposite directions, one at the rate of two miles in an hour, and the other six; how far apart are they in seven hours? How far in ten hours?

9. If the interest of one dollar is six cents a year, what is the interest of three dollars for two years?

10. How many oranges, at 4 cents apiece, must be given for 12 lemons, at 3 cents apiece?

11. How many pounds of beef, at 9 cents a pound, will pay for 6 dozen of eggs, at 12 cents a dozen?

12. If 8 men can do a certain piece of work in 6 days, in how many days can 4 men do the same?

13. If 9 men can do a piece of work in 8 days, how many men can do the same in 12 days?

14. If 4 men can do a piece of work in 6 days, in how many days can 3 men do twice as much work?

15. A farmer sold 2 tons of hay at $20 a ton, and for pay received 8 yards of cloth at $4 a yard, and the rest in money; how much money did he receive?

16. If a man earns $24 in 2 months, what will 2 men earn in 6 months?

17. Six times 8 are how many times 12?

18. Ten times 6 are how many times 12? 20? 5?

19. Eight times 8 are how many times 4? 32? 16?

20. Seven times 10 are how many times 5? 7? 14?

21. Seventy-two are how many times 8? 6? 9? 12? 4?

22. Eighty are how many times 4? 8? 20? 10? 5? 16?

23. Seventy-five are how many times 25? 5? 15? 3?

24. Eighty-four are how many times 12? 6? 7? 21? 4?

25. Ninety-six are how many times 8? 16? 12? 24? 48? 32?

26. One hundred are how many times 10? 20? 25? 50? 2? 4? 5?

27. One hundred and eight are how many times 12? 9? 54? 4? 27?

28. One hundred and twenty are how many times 12? 20? 10? 40?

29. One hundred and twenty-five are how many times 25? 5? 125? 1?

30. One hundred and thirty-two are how many times 11? 22? 44? 12? 6?

31. One hundred and forty-four are how many times 12? 24? 48? 6?

LESSON XI.

Read rapidly and accurately across the page.

9 and 1?	9 from 10?	9 times 1?	9 in 9?
9 and 2?	9 from 11?	9 times 2?	9 in 18?
9 and 3?	9 from 12?	9 times 3?	9 in 27?
9 and 4?	9 from 13?	9 times 4?	9 in 36?
9 and 5?	9 from 14?	9 times 5?	9 in 45?
9 and 6?	9 from 15?	9 times 6?	9 in 54?
9 and 7?	9 from 16?	9 times 7?	9 in 63?
9 and 8?	9 from 17?	9 times 8?	9 in 72?
9 and 9?	9 from 18?	9 times 9?	9 in 81?
9 and 10?	9 from 19?	9 times 10?	9 in 90?
9 and 11?	9 from 20?	9 times 11?	9 in 99?
9 and 12?	9 from 21?	9 times 12?	9 in 108?
10 and 1?	10 from 11?	10 times 1?	10 in 10?
10 and 2?	10 from 12?	10 times 2?	10 in 20?
10 and 3?	10 from 13?	10 times 3?	10 in 30?
10 and 4?	10 from 14?	10 times 4?	10 in 40?
10 and 5?	10 from 15?	10 times 5?	10 in 50?
10 and 6?	10 from 16?	10 times 6?	10 in 60?
10 and 7?	10 from 17?	10 times 7?	10 in 70?
10 and 8?	10 from 18?	10 times 8?	10 in 80?
10 and 9?	10 from 19?	10 times 9?	10 in 90?

10 and 10?	10 from 20?	10 times 10?	10 in 100?
10 and 11?	10 from 21?	10 times 11?	10 in 110?
10 and 12?	10 from 22?	10 times 12?	10 in 120?
11 and 1?	11 from 12?	11 times 1?	11 in 11?
11 and 2?	11 from 13?	11 times 2?	11 in 22?
11 and 3?	11 from 14?	11 times 3?	11 in 33?
11 and 4?	11 from 15?	11 times 4?	11 in 44?
11 and 5?	11 from 16?	11 times 5?	11 in 55?
11 and 6?	11 from 17?	11 times 6?	11 in 66?
11 and 7?	11 from 18?	11 times 7?	11 in 77?
11 and 8?	11 from 19?	11 times 8?	11 in 88?
11 and 9?	11 from 20?	11 times 9?	11 in 99?
11 and 10?	11 from 21?	11 times 10?	11 in 110?
11 and 11?	11 from 22?	11 times 11?	11 in 121?
11 and 12?	11 from 23?	11 times 12?	11 in 132?
12 and 1?	12 from 13?	12 times 1?	12 in 12?
12 and 2?	12 from 14?	12 times 2?	12 in 24?
12 and 3?	12 from 15?	12 times 3?	12 in 36?
12 and 4?	12 from 16?	12 times 4?	12 in 48?
12 and 5?	12 from 17?	12 times 5?	12 in 60?
12 and 6?	12 from 18?	12 times 6?	12 in 72?
12 and 7?	12 from 19?	12 times 7?	12 in 84?
12 and 8?	12 from 20?	12 times 8?	12 in 96?
12 and 9?	12 from 21?	12 times 9?	12 in 108?
12 and 10?	12 from 22?	12 times 10?	12 in 120?
12 and 11?	12 from 23?	12 times 11?	12 in 132?
12 and 12?	12 from 24?	12 times 12?	12 in 144?

LESSON XII.

1. FIFTY-ONE are how many times 10? 5? 3? 6? 9? 7? 8?

2. Fifty-four are how many times 9? 10? 6? 7? 8? 12? 11?

3. Fifty-two are how many times 4? 8? 12? 13? 10? 7? 6?

4. Sixty are how many times 6? 10? 8? 7? 12?

5. Fifty-six are how many times 7? 8? 9? 11? 12? 6? 5?

6. Fifty-three are how many times 7? 10? 9? 12? 4? 6?

7. Fifty-five are how many times 11? 10? 5? 6? 9? 8?

8. Fifty-eight are how many times 9? 6? 10? 8? 7? 5?

9. Fifty-seven are how many times 12? 11? 5? 10? 8? 6?

10. Fifty-nine are how many times 10? 5? 12? 8? 7? 9?

11. Sixty-four are how many times 8? 9? 10? 12? 6? 5?

12. Sixty-five are how many times 10? 5? 12? 9? 7? 8?

13. Sixty-two are how many times 12? 10? 9? 7? 8? 5?

14. Sixty-three are how many times 9? 7? 8? 5? 12? 6?

15. Sixty-one are how many times 12? 10? 8? 5? 6? 9?

16. Sixty-six are how many times 6? 11? 10? 9? 12? 8?

17. Sixty-nine are how many times 6? 12? 9? 8? 7? 11?

18. Sixty-seven are how many times 8? 7? 11? 12? 5? 9?

19. Sixty-eight are how many times 11? 12? 8? 7? 10? 6?

20. Seventy are how many times 10? 7? 12? 5?

21. Seventy-five are how many times 10? 5? 12? 6? 7? 9?

22. Doctor May has 72 miles to ride; how many hours will it take him, if he rides at the rate of 9 miles an hour? At the rate of 8 miles? 10 miles? 12 miles? 6 miles? 5 miles? 7 miles? 11 miles?

23. Seventy-two are how many times 9? 8? 10? 12? 6?

24. Seventy-six are how many times 12? 10? 6? 11? 5?

25. Seventy-eight are how many times 11? 7? 8? 10? 9?

26. Seventy-four are how many times 9? 10?. 8? 7? 12?

27. Seventy-seven are how many times 11? 7? 9? 8? 10?

28. Seventy-nine are how many times 11? 7? 8? 10? 9?

29. Seventy-one are how many times 10? 8? 7? 9? 6?

30. Seventy-three are how many times 7? 9? 10? 6? 12?

31. Eighty-four are how many times 7? 12? 10? 9? 11?

32. Ninety-six are how many times 12? 8? 9? 10? 6?

33. If you had $80, how many tons of hay could you buy at $10 a ton? How many at $12? At $8? At $7? At $6? At $9? At $11?

34. Eighty are how many times 10? 12? 8? 7? 6? 9? 11?

35. Ninety-two are how many times 10? 8? 9? 7? 12? 6?

36. Ninety-five are how many times 9? 10? 6? 8? 7? 11?

37. Eighty-one are how many times 8? 12? 10? 9? 7? 6?

38. Eighty-five are how many times 7? 10? 8? 12? 9? 11?

39. Ninety-one are how many times 9? 8? 10? 12? 7? 11?

40. Ninety are how many times 10? 12? 7? 9? 8? 6?

41. Ninety-eight are how many times 9? 8? 12? 7? 10? 11?

42. Eighty-two are how many times 8? 9? 10? 12? 11? 7?

43. One hundred are how many times 10? 9? 12? 11? 8?

44. Eighty-seven are how many times 7? 11? 12? 10? 8? 9?

45. Eighty-three are how many times 9? 8? 10? 12? 7? 11?

46. Eighty-nine are how many times 10? 8? 11? 12? 7? 9?

47. Ninety-nine are how many times 9? 11? 12? 8? 7? 10?

48. Ninety-four are how many times 12? 7? 10? 9? 8? 11?

49. Eighty-six are how many times 7? 10? 8? 9? 11? 12?

50. Ninety-seven are how many times 10? 9? 12? 8? 11?

51. Eighty-eight are how many times 10? 8? 12? 11? 7? 9?

LESSON XIII.

1. A FARMER has 30 hens in one yard, and 50 in another; how many hens has he in both yards?

Solution. 30 is 3 tens, and 50 is 5 tens; 3 tens and 5 tens are 8 tens, or 80; therefore, the farmer has 80 hens in the two yards.

2. How many are

	3 and 2?	30 and 20?	300 and 200?
3.	4 and 3?	40 and 30?	400 and 300?
4.	8 and 5?	80 and 50?	800 and 500?
5.	7 + 4?	70 + 40?	700 + 400?
6.	7 + 7?	70 + 70?	700 + 700?

7. John bought a knife for 45 cents, and a top for 23 cents; what did he pay for both?

Solution. 40 and 20 are 60, and 5 and 3 are 8; 60 and 8 are 68; therefore, he paid 68 cents for both.

8. How many are 48 and 35?

Solution. 40 and 30 are 70, and 8 and 5 are13; 70 and 13 are 83; therefore, 48 and 35 are 83.

9. How many are

	2 d 3?	12 and 13?	42 and 53?
10.	5 and 4?	25 and 34?	65 and 24?
11.	5 + ?	25 + 43?	55 + 73?
12.	8 + 4?	28 + 34?	68 + 74?

13. A farmer having 50 cows, sold 20 of them; how many had he remaining?

14. How many are

	5 less 2?	50 less 20?	500 less 200?
15.	7 less 3?	70 less 30?	700 less 300?
16.	6 — 2?	60 — 20?	600 — 200?
17.	9 — 5?	90 — 50?	900 — 500?

18. A farmer having $75, paid $32 for a colt; how many dollars had he left?

Solution. $75 are $70 and $5; $32 are $30 and $2; $70 less $30 are $40, and $5 less $2 are $3, and $40 and $3 are $43; therefore, he had $43 left.

19. A man having $53, paid $18 for a coat; how many dollars had he left?

Solution. $53 are $40 and $13; $18 are $10 and $8; $40 less $10 are $30, and $13 less $8 are $5; $30 and $5 are $35; therefore, etc.

20. How many are

	5 less 3?	65 less 23?	95 less 43?
21.	8 less 3?	38 less 13?	78 less 53?
22.	6 — 4?	56 — 24?	96 — 44?
23.	15 — 7?	65 — 27?	52 — 37?

24. At $40 per acre, what will 2 acres of land cost?

25. How many are

	2 times 4?	2 times 40?	2 times 400?
26.	3 times 2?	3 times 20?	3 times 200?
27	3 × 3?	30 × 3?	300 × 3?

28. At \$23 per acre, what will 6 acres of land cost?

Solution. 6 times \$20 are \$120, and 6 times \$3 are \$18; \$120 and \$18 are \$138; therefore, etc.

29. How many are

29.	5 times 3?	5 times 23?	5 times 63?
30.	6 times 2?	6 times 42?	6 times 62?
31.	8 × 2?	38 × 2?	68 × 2?
32.	7 × 3?	37 × 3?	57 × 3?

33. I have 3 equal fields which together contain 73 acres; how many acres are there in each field?

Solution. 73 equals 60 and 13; $\frac{1}{3}$ of 60 is 20, and $\frac{1}{3}$ of 13 is $4\frac{1}{3}$; 20 and $4\frac{1}{3}$ are $24\frac{1}{3}$; therefore, etc.

34.	What is $\frac{1}{2}$ of 6?	Of 60?	Of 600?
35.	What is $\frac{1}{2}$ of 8?	Of 28?	Of 58?
36.	What is $\frac{1}{3}$ of 8?	Of 38?	Of 98?
37.	How many are 8 ÷ 4?	28 ÷ 4?	88 ÷ 4?
38.	How many are 9 ÷ 4?	49 ÷ 4?	69 ÷ 4?

39. How many times

39.	3 in 6?	3 in 60?	3 in 600?
40.	4 in 8?	4 in 80?	4 in 800?
41.	5 in 8?	5 in 38?	5 in 78?

LESSON XIV

1. FOR 50 cents, how many dozen of eggs can be bought, at 25 cents a dozen? At 12 cents a dozen?

2. If you have 56 cents, how many pounds of sugar can you buy, at 7 cents a pound? At 8 cents? At 10 cents?

3. Two trains of cars, 64 miles apart, are running the same way; in how many hours will they be together, if the hindmost train gains upon the other 8 miles an hour? In how many hours if it gains 4 miles an hour? If 6 miles?

4. How many tons of hay, at \$12 a ton, can I buy for \$72? How many at \$10 a ton? At \$15?

5. How many dresses, containing 12 yards each,

can be made from 48 yards of silk? How many containing 16 yards each?

6. How many cords of wood, at $7 a cord, can be bought for $84? How many cords, at $8 a cord?

7. If there are 12 months in a year, how many years are there in 96 months? In 75 months?

8. If there are 30 days in a month, how many months are there in 90 days? In 60 days?

9. If there are 16 ounces in a pound, how many pounds are there in 80 ounces? How many if there are 12 ounces in a pound?

10. How many pounds of raisins, at 11 cents a pound, can you buy for 55 cents? How many at 10 cents a pound? At 15 cents?

11. Six men bought a houselot for $96; their shares being equal, what did each man pay?

12. Hired a horse and carriage for 77 cents, and rode 7 miles; what was the price per mile?

13. A man hired a horse and carriage for a ride, agreeing to pay 12 cents a mile; he paid 87 cents; how far did he ride?

14. Bought 15 cords of wood for $75; what was the price per cord?

15. Mr. Holt can do a piece of work in 78 hours; how many days will it take him to do it, if he works 12 hours each day? How many if he works 10 hours each day?

16. One man can do a piece of work in 90 hours; in how many hours can 6 men do it? 10 men? 12 men?

17. One man can do a piece of work in 90 hours; in how many hours can 6 men do half as much work? In how many hours can 10 men do twice as much work?

18. A dairy woman wishes to put 75 pounds of butter into 5 boxes; how many pounds shall she put into each box?

19. Eight cheeses weigh 100 pounds; what is their average weight?

20. If an orange is worth 6 apples, how many oranges are 66 apples worth?

21. If there are 20 shillings in 1 pound, how many pounds are there in 80 shillings? In 85 shillings?

22. How many miles in 96 furlongs, if there are 8 furlongs in 1 mile? How many miles in 82 furlongs? In 78 furlongs?

23. At 12 cents a pound, how many pounds of beef can be bought for 144 cents? For 93 cents?

24. How many wheelbarrows, at 7 dollars apiece, can be bought for 84 dollars? For 63 dollars?

25. A boy sold 6 doves for 75 cents; how many cents did he receive for 1 dove? For 3 doves? For 4 doves?

26. A boy sold chickens at 25 cents apiece, and received 1 dollar, or 100 cents, for them; how many chickens did he sell?

27. If wine is worth 20 cents a pint, how many pints may be bought for 1 dollar? How many quarts?

28. If brandy is worth 20 cents a gill, how many gills may be bought for 1 dollar? How many pints?

29. What is the price of a shawl, if 9 shawls cost $108?

30. A boy divided 80 chestnuts equally among 5 of his companions; how many did he give to each?

31. If 8 bushels of apples make 1 barrel of cider, how many barrels will 92 bushels make? 74 bushels? 78 bushels?

32. At $10 a barrel, how many barrels of flour can be bought for $85? For $92?

33. Eight times 10 are how many times 8?

34. Eight times 11 are how many times 12?

35. A man divided $84 equally among his 3 sons and 4 daughters; how many dollars did he give to each?

36. A boy having 7 melons, gave away 2 of them, and sold the rest for $1 ; what did he receive for each of those he sold?

37. A man having 75 dollars, bought 7 sheep, and had $5 left; what did he pay for each sheep?

38. A boy bought 5 hens at 20 cents each, and paid for them with apples at 10 cents a dozen; how many dozen did it take?

39. A man had 75 sheep, and bought 5 more; he then divided them equally in 8 pens; how many sheep did he put in each pen?

40. A boy had 50 peaches, and found 22 more; he then divided all of them equally among himself and 11 of his companions; how many did he give to each?

SECTION FOURTH.

LESSON I.

Remarks. 1. A single thing of any kind is called a *unit* or *one*. A unit is represented by the figure 1.

2. When any thing is divided into two or more equal parts, each of these parts is called a *fraction* of the thing which is divided.

3. A fraction, as well as a whole number, may be expressed by figures. Each fraction is usually expressed by means of two numbers, one written above the other, with a line between them; thus,

$\frac{1}{2}$,	. . . . . one half.	$\frac{1}{6}$,	. . . . . one sixth.
$\frac{1}{3}$,	. . . . . one third.	$\frac{5}{6}$,	. . . . . five sixths.
$\frac{2}{3}$,	. . . . . two thirds.	$\frac{6}{6}$,	. . . . . six sixths.
$\frac{1}{4}$,	. . . . . one fourth.	$\frac{3}{7}$,	. . . three sevenths.
$\frac{2}{4}$,	. . . . two fourths.	$\frac{9}{8}$,	. . . nine eighths.
$\frac{3}{4}$,	. . . three fourths.	$\frac{11}{12}$,	. . eleven twelfths.
$\frac{1}{5}$,	. . . . . . one fifth.	$\frac{19}{12}$,	. . nineteen twelfths.
$\frac{2}{5}$,	. . . . . two fifths.	$\frac{5}{21}$,	. . five twenty-firsts.

4. The number *below* the line shows *into how many parts a unit is divided.* It is called the DENOMINATOR, because it *denominates* or *gives name* to the parts; thus, if an apple, or any thing, is divided into *three* equal parts, each part is one *third* of the apple or thing; if into *eight*, each part is one *eighth.*

5. The number *above* the line is called the NUMERATOR, because it *numerates* or *numbers* the parts taken; thus the fraction $\frac{2}{3}$, indicates that a unit has been divided into *three* equal parts, and that *two* of the parts are used.

6. The numerator and the denominator are the TERMS of the fraction.

7. A PROPER FRACTION is one whose numerator is *less* than its denominator; as $\frac{1}{2}$, $\frac{3}{4}$, $\frac{7}{9}$. The value of a proper fraction is *less than one.*

8. An IMPROPER FRACTION is one whose numerator *equals* or *exceeds* its denominator; as, $\frac{4}{4}$, $\frac{6}{5}$, $\frac{9}{4}$. The value of an improper fraction is *one* or *more than one;* for this reason it is called an *improper* fraction.

9. A MIXED NUMBER is a whole number and a fraction united; as $2\frac{1}{2}$, read two and one half; $5\frac{2}{3}$, read five and two thirds.

LESSON II.

1. IN 3 apples how many halves of an apple?

Solution. There are 2 halves in 1 apple; therefore there are 3 times 2 halves, which are 6 halves, in 3 apples.

2. What fraction represents 6 halves? *Ans.* $\frac{6}{2}$.

3. A man divided 5 barrels of flour among his workmen, giving them $\frac{1}{2}$ of a barrel apiece; to how many men did he give the flour?

4. How many halves are there in 5 whole ones?

5. A man having 3 barrels and 1 half of a barrel of apples, wishes to give them to some poor people;

to how many can he give them, if he gives them $\frac{1}{2}$ of a barrel apiece?

6. How many halves are there in 3 and 1 half?

Solution. In 3 whole ones there are 3 times 2 halves, which are 6 halves; to this add the 1 half, and we have 7 halves or $\frac{7}{2}$; therefore, in 3 and 1 half there are 7 halves.

7. A man paid 1 half of a dollar apiece to several workmen; it took 4 dollars and 1 half of a dollar to pay them; how many men did he pay?

8. How many halves are there in 4 and 1 half; that is, in $4\frac{1}{2}$?

9. How many halves are there in 1? In 2? In 3? In 4? In 5?

10. How do you know how many halves there are in any number?

Ans. There are twice as many halves as there are whole ones, because it takes *two* halves to make *one* whole one.

11. In 6 melons how many half melons?

12. In $8 how many half dollars?

13. Change 9 to halves.

14. Change 6 and 1 half to halves.

15. Change $7\frac{1}{2}$ to halves.

16. Change $12\frac{1}{2}$ to halves.

17. If you cut 2 melons each into 3 equal pieces, how many pieces will you have? What part of a melon will each piece be?

18. If you cut 3 apples each into 3 equal pieces, how many pieces will they make? What part of an apple will each piece be?

19. If you cut 4 pies each into 3 equal pieces, how many pieces will they make?

20. How many thirds are there in 1? In 2? In 3? In 4? In 5? In 9?

21. How do you know how many thirds there are in any number?

Ans. There are 3 times as many thirds as there are whole ones, because it takes *three* thirds to make *one* whole one.

22. I have 1 cord and 1 third of a cord of wood to give to some poor persons; to how many can I give it, if I give them 1 third of a cord apiece?

23. How many thirds are there in $1\frac{1}{3}$?

24. A man gave 3 dollars and 2 thirds of a dollar to some poor persons, giving them 1 third of a dollar apiece; to how many persons did he give it?

25. How many thirds are there in $3\frac{2}{3}$?

26. How many thirds in $5\frac{1}{3}$? In $7\frac{2}{3}$? In $11\frac{2}{3}$? In $12\frac{1}{3}$?

27. If a boy eats 1 fourth of a pie at a meal, how many boys, at the same rate, will eat a whole pie at a meal? How many boys will eat 2 pies? 3 pies? 4 pies? 6 pies?

28. How many fourths are there in 1? In 2? 3? 4? 6? 8?

29. How do you know how many fourths there are in any number?

30. If a horse will eat 1 fourth of a bushel of oats in a day, in how many days will he eat a bushel? 2 bushels? 2 bushels and $\frac{3}{4}$ of a bushel? $3\frac{1}{4}$ bushels? $5\frac{3}{4}$ bushels?

31. How many fourths are there in 2 and 3 fourths? In $3\frac{1}{4}$? In $5\frac{3}{4}$? In $7\frac{1}{4}$? In $7\frac{3}{4}$?

32. If a boy can earn 1 fifth of a dollar in a day, in how many days can he earn 1 dollar? 2 dollars? 3 dollars? 6 dollars? 8 dollars?

33. How many fifths are there in 1? In 2? 3? 5? 10?

34. How do you know how many fifths there are in any number?

35. If a boy can earn 1 fifth of a dollar in a day, how many boys will it take to earn 1 dollar and 1 fifth of a dollar in the same time? How many to

earn 2 dollars and 3 fifths of a dollar? 4 dollars and 2 fifths of a dollar? $5\frac{4}{5}$ dollars? \$$7\frac{3}{5}$?

36. How many fifths are there in 1 and 1 fifth? In 2 and 3 fifths? In $4\frac{3}{5}$? In $5\frac{4}{5}$? In $8\frac{3}{5}$?

37. In 6 dollars and 2 fifths of a dollar how many fifths of a dollar?

38. In 6 and 2 fifths how many fifths?

39. In 7 and 3 fifths how many fifths?

40. In 9 and 4 fifths how many fifths?

41. If a quart of strawberries costs 1 sixth of a dollar, how many quarts can I buy for 1 dollar? For 2 dollars? 3 dollars? 5 dollars? 7 dollars?

42. How many sixths are there in 1? In 2? 3? 5? 8? 9?

43. How do you know how many sixths there are in any number?

44. In 2 and 1 sixth how many sixths?

45. In 3 and 5 sixths how many sixths?

46. In 5 and 3 sixths how many sixths?

47. In 6 and 4 sixths how many sixths?

48. How many sevenths are there in 1? In 2? 3? 5? 7? 9? 10?

49. How many sevenths in 1 and 3 sevenths?

50. How many sevenths in 1 and 6 sevenths?

51. How many sevenths in 2 and 4 sevenths?

52. How many sevenths in 3 and 3 sevenths?

53. How many sevenths in 5 and 2 sevenths?

54. How many sevenths in 9 and 6 sevenths?

55. How many eighths in 1? In 2? 3? 5? 7? 10? 12?

56. In 3 and 5 eighths how many eighths?

57. In 5 and 3 eighths how many eighths?

58. In 9 and 7 eighths how many eighths?

59. How many ninths are there in 1? In 2? 3? 5? 4? 8? 7?

60. In 2 and 5 ninths how many ninths?

61. In 3 and 8 ninths how many ninths?

62. In 5 and 4 ninths how many ninths?
63. In 7 and 7 ninths how many ninths?
64. How many tenths in 1? In 2? 3? 6? 9?
65. In 2 and 3 tenths how many tenths?
66. In 3 and 7 tenths how many tenths?
67. In 5 and 2 tenths how many tenths?
68. In 4 and 5 sixths how many sixths?
69. In 5 and 1 seventh how many sevenths?
70. In 9 and 2 fifths how many fifths?
71. In $4\frac{3}{8}$ how many times $\frac{1}{8}$?
72. In $5\frac{3}{10}$ how many times $\frac{1}{10}$?
73. In $2\frac{1}{12}$ how many times $\frac{1}{12}$?
74. In $3\frac{5}{11}$ how many elevenths?
75. In $6\frac{3}{5}$ how many fifths?

LESSON III.

1. If you give 2 boys 1 half of an apple apiece, how many apples will it take?

2. If you give 4 men 1 half of a dollar apiece, how many dollars will it take?

3. In 4 halves how many whole ones?

4. In 6 halves how many times 1?

5. In 8 halves how many times 1?

6. How do you know how many whole ones there are in any number of halves?

Ans. Since it takes *two* halves to make *one* whole one, there will be as many whole ones as 2 is contained times in the given number of halves; or, there will be half as many whole ones as there are halves.

7. What is one half of 8? One half of 10? Of 20? Of 24? Of 30? See page 48.

8. If you give 3 boys 1 half of a melon apiece, how many melons will it take?

9. What is 1 half of 3? *Ans.* Three halves $= 1\frac{1}{2}$.

10. If you give 5 poor people 1 half of a barrel of flour apiece, how many barrels will it take?

11. In 5 halves how many whole ones?

12. What is $\frac{1}{2}$ of 5? *Ans.* 5 halves $= 2\frac{1}{2}$.

13. If the interest of $1 is $\frac{1}{2}$ of a cent per month, what is the interest of the same sum for 7 months? For 8 months? For 9 months? For 11 months?

14. In 7 halves how many times 1?

15. In 9 halves how many times 1?

16. What is $\frac{1}{2}$ of 9? Of 11? 15? 21? 25?

17. A boy divided some oranges among 6 of his companions, giving 1 third of an orange to each; how many oranges did it take?

18. In 6 thirds how many whole ones?

19. What is $\frac{1}{3}$ of 6? *Ans.* 6 thirds $= 2$.

20. A teacher visited 9 different schools, spending 1 third of a day in each; how many days did it take?

21. In 9 thirds how many times 1?

22. What is $\frac{1}{3}$ of 9?

23. A man divided some coal among 7 poor families, giving 1 third of a ton to each family; how many tons did it take?

24. In 7 thirds how many times 1?

25. What is $\frac{1}{3}$ of 7?

26. A boy gave 1 third of a pineapple to each of his 12 playmates; how many pineapples did it take?

27. In 12 thirds how many times 1?

28. How do you know how many whole ones there are in any number of thirds?

29. What is $\frac{1}{3}$ of 12? Of 10? 15? 20?

30. If a man spends 1 fourth of a dollar in 1 day, how many dollars will he spend in 4 days? In 8 days? 12 days? 20 days?

31. In 8 fourths how many times 1?

32. In 20 fourths how many times 1?

33. How do you know how many whole ones there are in any number of fourths?

34. What is $\frac{1}{4}$ of 4? Of 8? Of 16? Of 24? Of 32? Of 40?

35. A benevolent lady gave 1 fourth of a dollar apiece to 9 poor children; how many dollars did it take?

36. In $\frac{9}{4}$ how many times 1? *Ans.* $2\frac{1}{4}$.

37. In 7 fourths how many times 1?

38. In 11 fourths how many times 1?

39. In 17 fourths how many times 1?

40. What is $\frac{1}{4}$ of 7? Of 11? Of 17? Of 23?

41. If 1 fifth of a ton of coal will supply a furnace 1 week, how many tons will supply it 10 weeks? 20 weeks? 15 weeks? 25 weeks?

42. In 10 fifths how many times 1?

43. In 8 fifths how many times 1?

44. In 12 fifths how many times 1?

45. In 19 fifths how many times 1?

46. If the interest of $1 is $\frac{1}{6}$ of a mill for 1 day, what is the interest of $1 for 6 days? For 18 days? For 9 days? For 20 days?

47. What is $\frac{1}{6}$ of 6? Of 18? Of 9? Of 20?

48. In 15 sixths how many times 1?

49. In 24 eighths how many times 1?

50. In 27 tenths how many times 1?

51. In 21 sevenths how many times 1?

52. In 48 ninths how many times 1?

53. In 45 fifths how many times 1?

54. In 27 twelfths how many times 1?

LESSON IV.

1. IF it takes 1 third of an hour to walk 1 mile, how long will it take to walk 2 miles?

2. What is 2 times 1 third?

3. If the interest of 1 dollar is 1 half of a cent for 1 month, what is the interest of 2 dollars for the same time? What the interest of 3 dollars? Of 4 dollars? Of 7 dollars? Of 10 dollars?

4. What is 2 times 1 half? 3 times 1 half? 4 times 1 half? 7 times $\frac{1}{2}$? 10 times $\frac{1}{2}$?

5. If 1 pound of coffee costs 1 fourth of a dollar, what will 3 pounds cost?

6. What is three times 1 fourth?

7. When potatoes are worth 3 fourths of a dollar a bushel, how many fourths of a dollar will 2 bushels cost? How many dollars?

8. Two times $\frac{3}{4}$ are how many fourths? How many times 1?

9. If a dinner for 1 man costs 2 thirds of a dollar, what will a dinner for 4 men cost?

10. Four times $\frac{2}{3}$ are how many thirds? How many times 1?

11. When eggs are worth 1 fifth of a dollar a dozen, what are 3 dozen worth?

12. What is 3 times 1 fifth?

13. Charles earned 2 fifths of a dollar on Monday; at the same rate per day how many fifths of a dollar does he earn in the week? How many dollars?

14. Six times $\frac{2}{5}$ are how many fifths? How many times 1?

15. If 1 bushel of oats costs 3 fifths of a dollar, what shall I pay for 7 bushels?

16. Seven times $\frac{3}{5}$ are how many fifths? How many times 1?

17. When corn is worth 4 fifths of a dollar a bushel, what will 6 bushels cost?

18. Six times 4 fifths are how many fifths? How many times 1?

19. If the interest of 1 dollar is 1 sixth of a mill for 1 day, what is the interest of 5 dollars for the same time? What the interest of 9 dollars?

20. What is 5 times $\frac{1}{6}$? 9 times $\frac{1}{6}$?

21. If the interest of $1 is 1 sixth of a mill per day, what is the interest of the same sum for 5 days? For 15 days? For 12 days?

22. If the interest of $1 for 1 day is 1 sixth of a mill, what is the interest of $3 for 4 days?

23. Four times 3 sevenths are how many sevenths? How many times 1?

24. Eight times 3 fourths are how many fourths? How many times 1?

25. What is 6 times 4 fifths?

26. What is 4 times 5 sixths?

27. What is 3 times 4 ninths?

28. What is 6 times 4 sevenths?

29. What is 9 times 3 eighths?

30. What is 5 times 9 tenths?

31. What is 8 times 3 elevenths?

32. What is 6 times $\frac{3}{8}$?

33. What is 4 times $\frac{3}{5}$?

34. What is 7 times $\frac{2}{3}$?

35. What is 9 times $\frac{3}{7}$?

36. What is 8 times $\frac{5}{6}$?

37. What is 5 times $\frac{4}{9}$?

LESSON V.

1. If corn is worth a dollar and a half a bushel, what shall I pay for 2 bushels?

2. When apples cost two dollars and a half a barrel, what will 3 barrels cost?

3. How many are 3 times 2 and 1 half?

Solution. Three times 2 are 6, and 3 times 1 half are 3 halves, or 1 and 1 half, which, added to 6, gives 7 and 1 half; therefore 3 times $2\frac{1}{2}$ are $7\frac{1}{2}$.

4. If a barrel of flour costs 6 dollars and a half, what will 5 barrels cost?

5. How many are 5 times 6 and 1 half?

6. What are 8 yards of cloth worth, at 3 dollars and a half per yard?

7. How many are 8 times 3 and 1 half?

8. If white lead is worth 9 dollars and 1 third per hundred weight, what are 2 hundred weight worth?

9. How many are twice 9 and 1 third?

10. When nails are worth 6 cents and 2 thirds per pound, what are 9 pounds worth?

11. How many are 9 times 6 and 2 thirds?

12. How many are 7 times 3 and 2 thirds?

13. How many are 8 times 4 and 1 third?

14. When flour costs 7 dollars and 3 fourths per barrel, what will 5 barrels cost?

15. How many are 5 times 7 and 3 fourths?

16. When apples cost 3 dollars and 1 fourth per barrel, what must I pay for 8 barrels?

17. How many are 8 times 3 and 1 fourth?

18. How many are 3 times 8 and 1 fourth?

19. How many are 5 times 11 and 3 fourths?

20. If raisins cost 3 dollars and 2 fifths per box, what will 2 boxes cost?

21. How many are 2 times 3 and 2 fifths?

22. How many are 5 times 4 and 3 fifths?

23. How many are 3 times 9 and 3 fifths?

24. If 1 man earns 1 dollar and 5 sixths in a day, how much will 8 men earn in the same time?

25. How many are 8 times 1 and 5 sixths?

26. How many are 9 times 4 and 1 sixth?

27. How many are 6 times 7 and 3 sixths?

28. If 1 man earns 2 dollars and 1 sixth in 1 day, how much will 3 men earn in 4 days?

29. If 12 yards and 3 sevenths of a yard of silk are required to make 1 dress, how many yards will be required for 5 dresses?

30. How many are 5 times 12 and 3 sevenths?

31. What cost 2 barrels of beef, at 18 dollars and 5 eighths per barrel?

32. How many are 2 times 18 and 5 eighths?

33. How many are 3 times 16 and 5 ninths?

34. How many are 4 times 20 and 3 eighths?

35. In an orchard of 10 trees, each tree bears 9 bushels and 5 eighths of a bushel of apples; how many bushels does the whole orchard bear?

36. If it takes 2 yards and 3 eighths of a yard of cloth to make a coat, and 1 yard and 5 eighths to make a pair of pantaloons, how many yards will it take to make 5 coats and 5 pair of pantaloons?

37. If 3 men can do a piece of work in 5 days and 4 ninths of a day, how many days will it take 1 man to do the same?

38. If 3 men can build 4 rods and 5 sixths of a rod of wall in 1 day, how many rods can they build in 5 days? In 4 days?

39. How many are 5 times $4\frac{5}{6}$?

Solution. Five times 4 are 20, and 5 times $\frac{5}{6}$ are $\frac{25}{6}$, or 4 and $\frac{1}{6}$, which, added to 20, make $24\frac{1}{6}$; therefore, 5 times $4\frac{5}{6}$ are $24\frac{1}{6}$.

40. How many are 4 times $4\frac{5}{6}$?

Solution. Four times 4 are 16, and 4 times $\frac{5}{6}$ are $\frac{20}{6}$, or $3\frac{1}{3}$, which, added to 16, make $19\frac{1}{3}$; therefore, 4 times $4\frac{5}{6}$ are $19\frac{1}{3}$?

41. How many are 4 times $2\frac{1}{3}$?

42. How many are 3 times $4\frac{3}{4}$?

43. How many are 5 times $3\frac{1}{4}$?

44. How many are 4 times $3\frac{1}{6}$?

45. How many are 4 times $3\frac{5}{6}$?

46. How many are 6 times $2\frac{5}{8}$?

47. How many are 9 times $5\frac{5}{6}$?

LESSON VI.

1. IF you divide a barrel of flour equally between 2 persons, what part of a barrel will each have?

2. If you divide a melon equally among 3 boys, what part of a melon will you give to each boy?

3. How can you divide 2 melons equally among 3 boys?

Ans. Divide each melon into 3 equal pieces, and then give each boy 1 piece from each melon?

4. What is 1 third of 2? *Ans.* 2 thirds of 1.

5. What is 1 third of 4? *Ans.* 4 thirds = $1\frac{1}{3}$.

6. What is 1 third of 5?

7. If 3 yards of silk cost 4 dollars, what is the price of 1 yard?

8. If 3 bushels of wheat cost $5, what is the price per bushel?

9. What is $\frac{1}{3}$ of 7? Of 8? Of 10? Of 11?

10. If a barrel of flour is divided equally among 4 persons, what part of a barrel does each receive? What would each receive if 2 barrels were divided? What if 3 barrels? If 5 barrels? If 6 barrels?

11. What is 1 fourth of 2? *Ans.* $\frac{2}{4}$, or $\frac{1}{2}$.

12. What is 1 fourth of 3?

Ans. 3 fourths of 1, or $\frac{3}{4}$.

13. What is 1 fourth of 5? Of 6? Of 7? Of 9?

14. If 5 boys eat a bushel of apples in a week, what part of a bushel will 1 boy eat in the same time? What will 2 boys eat? 3 boys? 4 boys? 6 boys?

15. What is $\frac{1}{5}$ of 2? *Ans.* $\frac{2}{5}$ of 1.

16. What is $\frac{1}{5}$ of 3? Of 4? 6? 9? 12?

17. What is $\frac{1}{6}$ of 2? Of 3? 4? 5? 8? 9? 10? 11? 15?

18. What is $\frac{1}{7}$ of 2? 3? 4? 5? 6? 8? 9? 13? 15?

19. What is $\frac{1}{8}$ of 2? 3? 4? 5? 6? 7? 9? 12?

20. What is $\frac{1}{9}$ of 2? 3? 4? 5? 6? 7? 8? 10? 11? 12?

21. What is $\frac{1}{10}$ of 2? 3? 4? 5? 8? 9? 10?

22. What is $\frac{1}{11}$ of 2? 6? 9? 12? 14? 15?

23. What is $\frac{1}{12}$ of 2? 3? 4? 5? 6? 7? 8? 9? 10? 11? 12? 15? 18? 20?

24. What is 2 thirds of 2?

Solution. One third of 2 is 2 thirds of 1, and *two* thirds of 2 are 2 times 2 thirds, which are 4 thirds, or 1 and 1 third; therefore, 2 thirds of 2 is $1\frac{1}{3}$.

25. If 1 barrel of apples cost $2, what will 1 third of a barrel cost? What 2 thirds of a barrel?

26. If a cord of wood costs \$5, what will 1 fourth of a cord cost? What will 3 fourths of a cord cost?

27. What is $\frac{3}{4}$ of 5? See Ex. 24.

28. If 4 bushels of corn cost \$3, what will 1 bushel cost? What will 3 bushels cost? 5 bushels?

29. What is $\frac{3}{4}$ of 3? $\frac{5}{4}$ of 3?

30. If a box of lemons costs \$4, what will $\frac{1}{5}$ of a box cost? $\frac{2}{5}$ of a box? $\frac{3}{5}$? $\frac{4}{5}$?

31. What is $\frac{1}{5}$ of 4? $\frac{2}{5}$ of 4? $\frac{3}{5}$? $\frac{4}{5}$?

32. If 7 gallons of molasses cost 5 dollars, what will 1 gallon cost? 2 gallons? 3 gallons? 5 gallons? 8 gallons?

33. What is $\frac{1}{7}$ of 5? $\frac{2}{7}$ of 5? $\frac{3}{7}$ of 5? $\frac{4}{7}$? $\frac{5}{7}$? $\frac{6}{7}$? $\frac{8}{7}$?

34. If 8 gallons of vinegar are worth 2 dollars, what is 1 gallon worth? What are 2 gallons worth? 3 gallons? 5 gallons? 7 gallons? 9 gallons?

35. What is $\frac{1}{8}$ of 2? $\frac{3}{8}$ of 2? $\frac{5}{8}$? $\frac{7}{8}$? $\frac{9}{8}$?

36. If 6 yards of satinet cost 5 dollars, what is the cost of 1 yard? Of 2 yards? 3 yards? 4 yards? 5 yards?

37. What is $\frac{1}{6}$ of 5? $\frac{2}{6}$ of 5? $\frac{3}{6}$ of 5? $\frac{5}{6}$? $\frac{7}{6}$? $\frac{9}{6}$?

38. What is $\frac{2}{3}$ of 8? $\frac{3}{4}$ of 9? $\frac{4}{3}$ of 7? $\frac{3}{8}$ of 11? $\frac{5}{6}$ of 10?

39. If 5 yards of cloth cost \$12, what will 1 yard cost? 2 yards? 3 yards? 4 yards? 6 yards? 8 yards?

40. What is $\frac{3}{5}$ of 17?

Solution. One fifth of 17 is $3\frac{2}{5}$, and *three* fifths are 3 times $3\frac{2}{5}$; 3 times 3 are 9, and 3 times $\frac{2}{5}$ are $\frac{6}{5}$, or $1\frac{1}{5}$, which, added to 9, makes $10\frac{1}{5}$; therefore, etc.

NOTE. This method is best for large numbers, and that given in Ex. 24 is best for small numbers.

41. What is 7 fifths of 17?

42. What is 3 sevenths of 24?

43. What is 7 ninths of 20?

44. What is $\frac{5}{6}$ of 19? $\frac{3}{8}$ of 21? $\frac{4}{9}$ of 26?

45. What is $\frac{5}{7}$ of 15? $\frac{3}{10}$ of 18? $\frac{5}{11}$ of 26?
46. What is $\frac{5}{12}$ of 17? $\frac{7}{12}$ of 19? $\frac{11}{6}$ of 13?
47. What is $\frac{3}{8}$ of 24? $\frac{5}{9}$ of 36? $\frac{7}{3}$ of 24?
48. What is $\frac{3}{11}$ of 44? $\frac{5}{7}$ of 42? $\frac{9}{11}$ of 77?
49. What is $\frac{5}{9}$ of 27? $\frac{3}{7}$ of 49? $\frac{5}{9}$ of 81?
50. What is $\frac{5}{3}$ of 24? $\frac{7}{6}$ of 42? $\frac{3}{8}$ of 80?
51. What is $\frac{7}{11}$ of 33? $\frac{6}{5}$ of 40? $\frac{7}{9}$ of 63?
52. What is $\frac{5}{8}$ of 64? $\frac{3}{7}$ of 63? $\frac{9}{12}$ of 60?

LESSON VII.

1. IF 1 half of a barrel of apples costs 1 dollar and 1 third of a dollar, what will a barrel cost?

2. 1 and 1 third is 1 half of what number?

Solution. Since 1 half of the number is 1 and 1 third, 2 halves, or the whole of the number, will be 2 times 1 and 1 third, which are 2 and 2 thirds; therefore, 1 and 1 third is 1 half of 2 and 2 thirds.

3. If 1 third of a box of oranges costs 2 dollars and 1 fourth, what will 2 thirds of a box cost? What a whole box?

4. 2 and 1 fourth is 1 third of what number?

5. If 1 fourth of a ton of hay costs 3 dollars and 1 half, what will 2 fourths of a ton cost? What will 3 fourths of a ton cost? What a whole ton?

6. 3 and 1 half is 1 fourth of what number?

7. If 1 fifth of a ton of coal costs 1 dollar and 3 fourths, what will 2 fifths of a ton cost? What 3 fifths? 4 fifths? 5 fifths, or a whole ton?

8. 1 and 3 fourths is 1 fifth of what number?

9. If 1 sixth of a barrel of flour costs 1 dollar and 3 fifths, what will 2 sixths of a barrel cost? 3 sixths? 4 sixths? 5 sixths? 6 sixths, or a whole barrel?

10. 1 and 3 fifths is 1 sixth of what number?

11. If a man can walk 4 miles and 5 sixths of a mile in 1 seventh of a day, how far can he walk in 2

sevenths of a day? In 3 sevenths? In 4 sevenths? In a whole day?

12. 4 and 5 sixths is 1 seventh of what number?

13. A boy skated 1 mile and 5 sevenths in 1 eighth of an hour; how far, at the same rate, would he skate in 2 eighths of an hour? In 3 eighths? In 5 eighths? In an hour?

14. 1 and 5 sevenths is 1 eighth of what number?

15. A locomotive ran 3 miles and 2 fifths of a mile in 1 ninth of an hour; how far would it run in 2 ninths of an hour? How far in 3 ninths of an hour? In 4 ninths? In 6 ninths? In an hour?

16. 3 and 2 fifths is 1 ninth of what number?

17. A ship sailed 1 mile and 1 fourth in 1 tenth of an hour; how far would she sail in 2 tenths of an hour? How far would she sail in 3 tenths of an hour? In 4 tenths? In 5 tenths? In 9 tenths? In an hour?

18. 1 and 1 fourth is 1 tenth of what number?

19. 6 and 2 fifths is 1 seventh of what number?

20. 4 and 3 sevenths is 1 fifth of what number?

21. 5 and 3 fourths is 1 eighth of what number?

22. 6 and 5 eighths is 1 fourth of what number?

23. 3 and $\frac{5}{6}$ is $\frac{1}{9}$ of what number?

24. 4 and $\frac{2}{3}$ is $\frac{1}{10}$ of what number?

25. 3 and $\frac{5}{9}$ is $\frac{1}{6}$ of what number?

26. $5\frac{4}{7}$ is $\frac{1}{2}$ of what number?

27. $2\frac{1}{5}$ is $\frac{1}{12}$ of what number?

28. $3\frac{2}{3}$ is $\frac{1}{11}$ of what number?

29. If 2 thirds of a box of lemons cost 3 dollars, what will one third of a box cost? What will a box cost?

30. 3 is 2 thirds of what number?

Solution. Since 3 is *two* thirds, *one* third is 1 half of 3, which is 1 and 1 half; and *three* thirds, or the whole, will be three times 1 and 1 half, which are 4 and 1 half; therefore, 3 is 2 thirds of 4 and 1 half.

31. 5 is 2 thirds of what number?

32. If 3 fourths of a yard of cloth cost $4, what will 1 fourth of a yard cost? What will a yard cost?

33. If 4 is 3 fourths of some number, what is 1 fourth of the same number? 1 and 1 third is 1 fourth of what number? Then 4 is 3 fourths of what number?

34. If $\frac{2}{5}$ of a box of lemons cost $3, what will a box cost?

35. If 3 is $\frac{2}{5}$ of some number, what is $\frac{1}{5}$ of the same number? $1\frac{1}{2}$ is $\frac{1}{5}$ of what number? Then 3 is $\frac{2}{5}$ of what number?

36. If $\frac{5}{6}$ of a hundred weight of sugar cost $8, what will 1 hundred weight cost?

37. If 8 is $\frac{5}{6}$ of some number, what is $\frac{1}{6}$ of the same number? $1\frac{3}{5}$ is $\frac{1}{6}$ of what number? Then 8 is $\frac{5}{6}$ of what number?

38. If $\frac{5}{8}$ of a cord of wood cost $4, what will a cord cost?

39. 4 is $\frac{5}{8}$ of what number?

40. If $\frac{7}{10}$ of a ton of hay cost $16, what will a ton cost?

41. 16 is $\frac{7}{10}$ of what number?

42. If a man can earn $23 in $\frac{5}{12}$ of a month, what can he earn in a month?

43. 23 is $\frac{5}{12}$ of what number?

44. A man can do $\frac{2}{9}$ of a piece of work in 5 days; how long will it take him to do the whole of it?

45. 5 is $\frac{2}{9}$ of what number?

46. A horse traveled 4 miles in $\frac{5}{9}$ of an hour; how far, at the same rate, would he travel in an hour?

47. 4 is $\frac{5}{9}$ of what number?

48. If 7 pounds of flour cost 38 cents, what will 12 pounds cost?

49. 38 is $\frac{7}{12}$ of what number?

50. If 6 yards of cloth costs $10, what will 5 yards cost? 8 yards?

51. If 3 men can do a piece of work in 7 hours, how long will it take 5 men to do the same work?

52. If 2 pipes will empty a cistern in 7 hours, in what time will 5 pipes empty it? 3 pipes?

53. 7 is $\frac{5}{2}$ of what number?

54. 9 is $\frac{4}{9}$ of what number?

55. 18 is $\frac{7}{5}$ of what number?

56. 23 is $\frac{6}{7}$ of what number?

57. A man bought 1 barrel and 1 fourth of a barrel of oil for $32; what was the price per barrel?

Solution. A barrel and a fourth of a barrel are the same as 5 fourths of a barrel. Since 5 fourths of a barrel cost 32 dollars, 1 fourth costs 1 fifth of 32 dollars, which is 6 dollars and 2 fifths, and 4 fourths cost 4 times 6 dollars and 2 fifths, which are 25 dollars and 3 fifths of a dollar; therefore, etc.

58. A man bought 1 acre and 3 fourths of an acre of land for $75; what was the price per acre?

59. 75 is $\frac{7}{4}$ of what number?

60. Bought $2\frac{3}{4}$ yards of cloth for $12; what was the price per yard?

61. 12 is $\frac{11}{4}$ of what number?

62. Bought $1\frac{3}{5}$ tons of hay for $30; what was the price of 1 ton?

63. Paid $18 for $1\frac{5}{6}$ barrels of cranberries, what did I pay for 1 barrel?

64. Paid $7 for $1\frac{5}{7}$ boxes of raisins; what, at the same rate, should I pay for 2 boxes?

65. Twice $\frac{7}{12}$ of 7 are how many?

66. Paid $3 for $1\frac{3}{8}$ bushels of wheat; what would 3 bushels cost?

67. 3 times $\frac{8}{11}$ of 3 are how many?

68. Bought $1\frac{1}{10}$ barrels of apples for $4; what should I pay for 4 barrels?

69. 4 times $\frac{10}{11}$ of 4 are how many?

70. Bought $1\frac{4}{11}$ gallons of vinegar for 33 cents; what would be the cost of 2 gallons?

71. 2 times $\frac{11}{15}$ of 33 are how many?

72. 5 times $\frac{4}{3}$ of 21 are how many?

SECTION FIFTH.

LESSON I.

UNITED STATES MONEY, sometimes called *Federal Money*, is the Currency of the United States.

TABLE.

10 Mills (*m.*)	make 1 Cent,	marked c.
10 Cents	" 1 Dime,	" d.
10 Dimes	" 1 Dollar,	" $.
10 Dollars	" 1 Eagle,	" e.

NOTE. The words *eagle* and *dime* are seldom or never used in business transactions. Dollars and cents are usually separated by a *point*; thus, $36.42 is read, 36 dollars and 42 cents.

TABLE OF ALIQUOT OR EXACT PARTS OF A DOLLAR.

50 cents = $\frac{1}{2}$ of a dollar,	20 cents = $\frac{1}{5}$ of a dollar,
$33\frac{1}{3}$ cents = $\frac{1}{3}$ of a dollar,	$16\frac{2}{3}$ cents = $\frac{1}{6}$ of a dollar,
25 cents = $\frac{1}{4}$ of a dollar,	$12\frac{1}{2}$ cents = $\frac{1}{8}$ of a dollar.

1. How many cents in 2 dimes? In 3 dimes and 6 cents?

2. How many cents in 30 mills? In 45 mills?

3. Change $4 and 6 dimes to dimes. $1, 2 dimes, and 5 cents to cents.

4. How many cents in $\frac{1}{2}$ of a dollar? In $\frac{1}{4}$? $\frac{1}{3}$? $\frac{1}{5}$? $\frac{1}{8}$? $\frac{1}{6}$? $\frac{1}{16}$?

5. What will 6 bushels of apples cost, at $33\frac{1}{3}$ cents per bushel?

Solution. Since 1 bushel costs 1 third of a dollar, 6 bushels will cost 6 times 1 third of a dollar, which are 6 thirds of a dollar, or 2 dollars.

6. At $16\frac{2}{3}$ cents per pound, what will 9 pounds of raisins cost?

7. At 25 cents per pound, what will 20 pounds of coffee cost?

8. At $12\frac{1}{2}$ cents a pound, what will 24 pounds of sugar cost?

9. Bought 12 sheep, at $$3.33\frac{1}{3}$ per head; what did they cost?

10. Bought 5 barrels of flour, at $9.33⅓ per barrel; what did I pay for the lot?

11. Paid 6¼ cents a pound for 12 pounds of nails; what did they cost?

12. At 37½ cents a yard, what will 8 yards of ribbon cost?

ENGLISH OR STERLING MONEY is the currency of Great Britain.

TABLE.

4 Farthings (*qr.*)	make 1 Penny,	marked d.
12 Pence	" 1 Shilling,	" s.
20 Shillings	" 1 Pound,	" £.

NOTE. A Sovereign is a gold coin equal to a pound. A Guinea is 21 shillings.

13. How many farthings in 2d.? In 3d. 2qr.?

14. How many pence in 3s.? In 2s. 9d.?

15. How many pounds in 46s.? In 65s.? In 82s. 6d.?

16. Change 1£ 2s. 4d. to pence. 3£ 5s. 6d. to shillings.

17. At 5s. a yard, how many pounds will 12 yards of cloth cost? 18 yards?

18. At 6d. a pound, how many shillings will 18 pounds of cheese cost? 24 pounds? 27 pounds?

19. At 8s. a pair, how many pair of shoes may be bought for 2£? For 3£ 4s.?

LESSON II.

TROY WEIGHT is used in weighing gold, silver, and precious stones.

TABLE.

24 Grains (*gr.*)	make 1 Pennyweight,	marked dwt
20 Pennyweights	" 1 Ounce,	" oz.
12 Ounces	" 1 Pound,	" lb.

1. How many grains in 2dwt.? In 2dwt. 6gr.?
2. How many grains in 2½dwt.? In 1¾dwt.?
3. How many pennyweights in 48gr.? In 72gr.?
4. How many pennyweights in 36gr.? In 42gr.?

5. How many pounds in 24oz.? In 72oz.? In 30oz.? In 51oz.? In 69oz.?

6. How many ounces in 3 lb.? In 2 lb. 3oz.?

7. How many dollars and cents will a gold chain weighing 12dwt. cost, at 9 dimes per pennyweight?

APOTHECARIES' WEIGHT is used in *mixing* medicines.

TABLE.

20 Grains (*gr.*)	make 1 Scruple,	marked sc., or ℈.
3 Scruples	" 1 Dram,	" dr., or ʒ.
8 Drams	" 1 Ounce,	" oz., or ℥.
12 Ounces	" 1 Pound,	" lb., or ℔

NOTE. The pound, ounce, and grain are the same in Apothecaries' Weight that they are in Troy Weight; but the ounce is differently divided.

8. How many grains in 2℈? In 3℈ 5gr.?

9. How many scruples in 40gr.? In 25gr.?

10. How many scruples in 5dr.? In 6dr. 2sc.?

11. How many ounces in 40dr.? In 46dr.?

12. In mixing a certain medicine, an apothecary used 10gr. of one kind, 2℈ of another kind, and 3ʒ of another; how many grains did he use in all?

AVOIRDUPOIS WEIGHT is used in weighing the coarser articles of merchandise; such as hay, cotton, tea, sugar, copper, iron, etc.

TABLE.

16 Drams (*dr.*)	make 1 Ounce,	marked oz.
16 Ounces	" 1 Pound,	" lb.
25 Pounds	" 1 Quarter,	" qr.
4 Quarters	" 1 Hundred Weight,	" cwt.
20 Hundred Weight	" 1 Ton,	" t.

13. At 9 cents a pound, what will 1cwt. of sugar cost?

14. Bought 3t. 10cwt. of hay, at $12 a ton; what was the whole cost?

15. At $12\frac{1}{2}$ cents a pound, what will 12 lb. of beef cost?

16. At $16\frac{2}{3}$ cents a pound, what will 15 lb. of butter cost?

17. What cost 24 lb. of honey, at $33\frac{1}{3}$ cents a pound?

LESSON III.

CLOTH MEASURE is used in measuring cloth, ribbons, braids, etc.

TABLE.

$2\frac{1}{4}$ Inches (*in.*)	make 1 Nail,	marked	na.
4 Nails	" 1 Quarter,	"	qr.
4 Quarters	" 1 Yard,	"	yd.

1. At 4c. a nail, what will 1yd. of ribbon cost?

2. At 3c. an inch, what will 1qr. of a yard of cloth cost?

3. At $1\frac{1}{2}$ a quarter, what will $3\frac{1}{2}$yd. of cloth cost?

4. If 2yd. 3qr. 1na. of cloth are required for 1 coat, how many yards are required for 4 coats?

5. What cost 3yd. 1qr. 2na. of velvet, at $2 a quarter?

6. How many inches in 1yd. 3qr. 2na. 1in.?

LONG MEASURE is used in measuring lengths or distances.

TABLE.

12 Inches (*in.*)	make 1 Foot,	marked	ft.
3 Feet	" 1 Yard,	"	yd.
$5\frac{1}{2}$ Yards, or $16\frac{1}{2}$ Feet	" 1 Rod,	"	rd.
40 Rods	" 1 Furlong,	"	fur.
8 Furlongs	" 1 Mile,	"	m.
3 Miles	" 1 League,	"	l.
60 Geographical, or $69\frac{1}{2}$ Statute Miles, nearly	" 1 Degree on the Circumference of the Earth,		1°.
360 Degrees	" 1 Circumference,		circ.

7. How many rods in 1 mile?

8. What is the cost of making 40rd. of road, at $1\frac{1}{2}$ a rod?

9. How many leagues in 15 miles?

10. If a ship sail 20 leagues a day, in how many days will she sail 300 miles?

11. In 3 rods how many yards? Feet?

12. If a man can travel 4 miles in an hour, how long will it take him to travel 3 leagues?

13. Hired a horse and carriage, at $16\frac{2}{3}$ cents per mile; what did I pay for a ride of 15 miles?

14. In 33 ft. how many yards? Rods?

LESSON IV.

Square Measure is used for measuring surfaces.

TABLE.

144	Square Inches (*sq. in.*)	make	1 Square Foot,	marked	sq. ft.
9	Square Feet	"	1 Square Yard,	"	sq. yd.
$30\frac{1}{4}$ / $272\frac{1}{4}$	Square Yards, or / Square Feet	"	1 Square Rod,	"	sq. rd.
40	Square Rods	"	1 Rood,	"	r.
4	Roods	"	1 Acre,	"	a.
640	Acres	"	1 Square Mile,	"	sq. m.

A Fig. 1. B

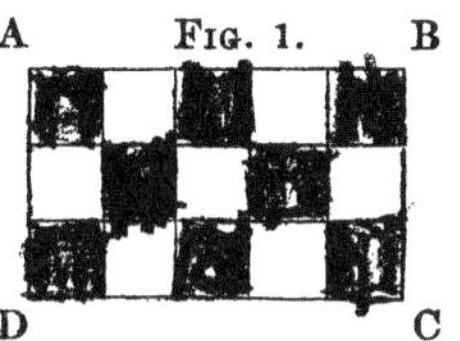

D C

Remarks. 1. A rectangle is a four-sided figure, which has its *corners* or *angles* all equal to each other, as ABCD, Fig. 1.

A Fig. 2. B

D C

2. A *square* is a *rectangle* whose *sides* are all equal to each other, as ABCD, Fig. 2.

3. The small checks in Fig. 1 are squares.

Ex. 1. A certain rectangular garden is 5 rods long and 3 rods wide; how many square rods does it contain? See Fig. 1.

2. How many rods round a rectangular garden which is 5 rods long and 3 rods wide? See Fig. 1.

3. How many square rods in a garden that is 6 rods square?

4. How many rods round a square garden, each side being 6 rods long?

5. How many square rods in a rectangular garden which is 10 rods long and 6 rods wide? How many rods round this garden? What will it cost to fence it, the fence costing $1.25 a rod?

6. At $2 per square rod, what cost 2 roods of land?

7. How many square feet in a board 10ft. long and 1ft. 6in. wide? 12ft. long and 2ft. 3in. wide?

8. I have a rectangular garden, 10 rods long and containing 50 square rods; how wide is it?

SOLID OR CUBIC MEASURE is used in measuring things which have length, breadth, and thickness.

TABLE.

1728 Cubic Inches (*c. in.*)	make 1 Cubic Foot,	marked cu. ft.
27 Cubic Feet	" 1 Cubic Yard,	" c. yd.
16 Cubic Feet	" 1 Cord Foot,	" c. ft.
8 Cord Feet, or 128 Cubic Feet	" 1 Cord,	" c.

FIG. 3

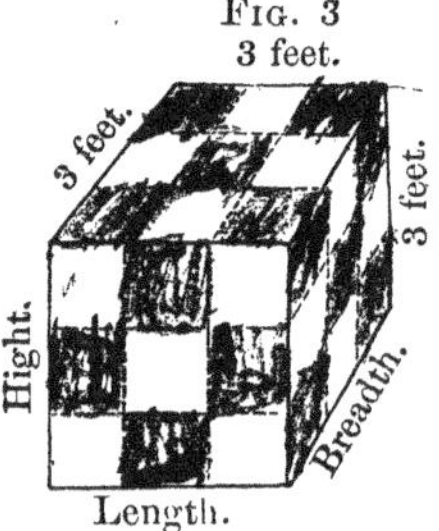

REMARKS. 1. A *cube* is a body, like Fig. 3, which is bounded by 6 *equal square faces.*

2. A cube is also called a *rectangular prism.* A rectangular prism may have its length, breadth, and hight *unequal.*

3. The *volume* or *contents* of a cube or rectangular prism may be found by multiplying the length by the breadth, which gives the area of the top or bottom face, and then multiplying this product by the hight.

Ex. 9. How many square feet are there in the top surface of Fig. 3?

10. How many cubic feet are there in Fig. 3?

11. How many square feet are there in the 6 faces of Fig. 3?

12. How many cubic inches in a rectangular prism or block which is 4 inches long, 3 inches wide, and 2 inches thick? How many square inches in one of its largest faces? How many in one of its smallest faces? How many in one of the other faces? How many in its 6 faces?

13. How many square feet in the surface of a cubical box whose edges are 2 feet in length?

14. How many cubic feet in a cube, each edge of which is 4 feet long?

15. How many cord feet in 48 cubic feet of wood?

16. How many cord feet in 3 cords of wood?

17. What cost 3 cords of wood, if 4 cord feet cost $3?

18. How many cubic feet in 3c. ft. 8cu. ft?

LESSON V.

LIQUID MEASURE is used in measuring water, milk, beer, wine, molasses, oil, etc.

TABLE.

4 Gills (*gi.*)	make 1 Pint,	marked pt.
2 Pints	" 1 Quart,	" qt
4 Quarts	" 1 Gallon,	" gal.

1. At 4c. a quart, what cost 2gal. 3qt. of milk?

2. At 25c. a pint, what will 1gal. 2qt. 1pt. of wine cost?

3. At 60 cents a gallon, what will 2gal. 2qt. 1pt. of oil cost?

4. What cost 2gal. 3qt. 1pt. of vinegar, at 4c. a quart?

5. What cost 2qt. 1pt. 3gi. of molasses, at 2c. a gill?

DRY MEASURE is used in measuring grain, fruit, potatoes, salt, charcoal, etc.

TABLE.

2 Pints (*pt.*)	make 1 Quart,	marked qt.
8 Quarts	" 1 Peck,	" pk.
4 Pecks	" 1 Bushel.	" bush.

6. At 10c. a quart, what costs 1 bushel of cherries?

7. At $1 a bushel, what cost 3pk. of corn?

8. If 1 bushel of chestnuts costs $2, what will 6qt. cost?

9. Bought a bushel of hickory nuts for $1.60 and sold them at $6\frac{1}{4}$c. a quart; how much did I gain?

LESSON VI.

Time is used in measuring duration.

TABLE.

60	Seconds (*sec.*)	make	1 Minute,	marked m.
60	Minutes	"	1 Hour,	" h.
24	Hours	"	1 Day,	" d.
7	Days	"	1 Week,	" wk.
52 / 365¼	Weeks and 1¼ Days, or Days	"	1 Year,	" yr.
100	Years	"	1 Century,	" c.

The names of the seasons and of the months, and the number of days in the several months, are as follows:

Seasons.		Months.	No. of Days.
Winter,	1st.	January,	31
	2d.	February,	28, in leap-year 29.
Spring,	3d.	March,	31
	4th.	April,	30
	5th.	May,	31
Summer,	6th.	June,	30
	7th.	July,	31
	8th.	August,	31
Autumn,	9th.	September,	30
	10th.	October,	31
	11th.	November,	30
Winter,	12th.	December,	31

The number of days in each month may be easily remembered by committing the following lines

"Thirty days hath September,
April, June, and November;
All the rest have thirty-one
Save the second month alone,
Which has just eight and a score
Till leap-year gives it one more."

1. How many seconds in 1m. 30sec.?

2. How many minutes between half-past 9 o'clock and noon?

3. If a man earns $1½ per day, how many dollars will he earn in the 6 working-days of a week?

4. How many days in leap year from the 1st of January to the 10th of March, inclusive? How many in a common year?

5. How many days in June, July, and August?

6. If a boy can count 3 in a second, how many can he count in a minute?

7. John slept 8 hours each night; how many hours did he sleep in a week?

8. How many hours in 3½ days? In 2d. 6h.?

MISCELLANEOUS TABLE.

12	Single things	make	1 Dozen.
12	Dozen	"	1 Gross.
12	Gross	"	1 Great Gross.
20	Single things	"	1 Score.
5	Score	"	1 Hundred.
24	Sheets of paper	"	1 Quire.
20	Quires	"	1 Ream.
250	sheets of paper	"	1 Printer's Token (small).
500	" " "	"	1 " " (large).
56	Pounds	"	1 Bushel of corn.
60	Pounds	"	1 Bushel of wheat.
196	Pounds	"	1 Barrel of flour.
200	Pounds	"	1 Barrel of beef, pork, or fish.

9. At 10c. a dozen, what cost 2 gross of buttons?

10. At 3c. apiece, what cost 2 dozen of oranges?

11. What cost 2 reams of paper, at 10c. a quire?

12. What cost 2 barrels of pork, at $8 per cwt.?

SECTION SIXTH.

LESSON I.

1. AT 2 cents a yard, what will 3 yards and one half of a yard of braid cost?

Solution. Three yards will cost 3 times 2 cents, which are 6 cents; one half of a yard will cost one half of 2 cents, which is 1 cent; and 1 cent added to 6 cents makes 7 cents; therefore, etc.

2. 3 times 2 and one half of 2 are how many?

3. At 3 dollars a barrel, what will 2 barrels and one third of a barrel of apples cost?

4. 2 times 3 and one third of 3 are how many?

5. At 3 dollars a yard, what will 4 yards and 2 thirds of a yard of broadcloth cost?

6. 4 times 3 and 2 thirds of 3 are how many?

Solution. 4 times 3 are 12, and 2 thirds of 3 are 2, which, added to 12, make 14; therefore, etc.

7. If a man walks 4 miles per hour, how far will he walk in 3 hours and one fourth of an hour?

8. 3 times 4 and one fourth of 4 are how many?

9. If a man earns 4 shillings per day, how ma shillings will he earn in 5 days and 3 fourths of a

10. 5 times 4 and 3 fourths of 4 are how man

11. If a man earns $5 per week, how many d lars will he earn in 4 weeks and one fifth of a week? How many in 6 weeks and 2 fifths of a week?

12. 4 times 5 and 1 fifth of 5 are how many?

13. 6 times 5 and 2 fifths of 5 are how many?

14. 5 times 5 and 3 fifths of 5 are how many?

15. 7 times 5 and 4 fifths of 5 are how many?

16. If wood is worth $6 a cord, what will 3 cords and 1 sixth of a cord cost? What will 7 cords and 5 sixths of a cord cost?

17. 3 times 6 and 1 sixth of 6 are how many?

18. 7 times 6 and 5 sixths of 6 are how many?

19. At $7 a ton, what will 4 tons and 1 seventh of a ton of coal cost? What will 6 tons and 3 sevenths of a ton cost?

20. 4 times 7 and 1 seventh of 7 are how many?

21. 6 times 7 and 3 sevenths of 7 are how many?

22. 7 times 7 and 5 sevenths of 7 are how many?

23. At 8 dollars a barrel, what will 3 barrels and 1 eighth of a barrel of flour cost? What will 7 barrels and 5 eighths of a barrel cost?

24. 3 times 8 and 1 eighth of 8 are how many?

25. 7 times 8 and 5 eighths of 8 are how many?

26. 4 times 6 and 2 sixths of 6 are how many?

27. 8 times 7 and 2 sevenths of 7 are how many?

28. 4 times 7 and 6 sevenths of 7 are how many?

29. 7 times 8 and 3 eighths of 8 are how many?

30. 9 times 8 and 7 eighths of 8 are how many?

31. If a pound of pork costs 9 cents, what will 3 pounds and 1 ninth of a pound cost? What will 5 pounds and 4 ninths of a pound cost?

32. 3 times 9 and 1 ninth of 9 are how many?

33. 5 times 9 and 4 ninths of 9 are how many?

34. 6 times 9 and 7 ninths of nine are how many?

35. If a ton of hay is worth $10, what are 4 tons and 3 tenths of a ton worth? What are 6 tons and 7 tenths of a ton worth?

36. 4 times 10 and 3 tenths of 10 are how many?

37. 6 times 10 and 7 tenths of 10 are how many?

38. 2 times 10 and 1 tenth of 10 are how many?

39. 5 times 10 and 9 tenths of 10 are how many?

40. 3 times 10 and 6 tenths of 10 are how many?

41. At 11 dollars an acre, what will 3 acres and 1 eleventh of an acre of land cost?

42. What is meant by one eleventh of any number? What by two elevenths, three elevenths, etc.?

43. 3 times 11 and 1 eleventh of 11 are how many?

44. 6 times 11 and 4 elevenths of 11 are how many?

45. If a car runs 12 miles per hour, how far will it run in 4 hours and 5 twelfths of an hour?

46. 4 times 12 and 5 twelfths of 12 are how many?

47. 3 times 12 and 7 twelfths of 12 are how many?

LESSON II.

REMARK. Six shillings make a dollar in many of the United States, and in this book it will be so understood, unless otherwise mentioned.

1. A boy bought 2 pencils, at 6 cents apiece; how many cents did they cost? He paid for them with peaches, at 3 cents apiece; how many peaches did it take?

2. 2 times 6 are how many times 3?

3. A man bought 4 barrels of apples, at $3 a bar-

rel; how many dollars did they cost? How much wood, at $6 a cord, would it take to pay for them?

4. 4 times 3 are how many times 6?

5. Mary bought 4 oranges, at 3 cents apiece; how many cents did they cost? She paid for them with pears, at 2 cents apiece; how many pears did it take?

6. 4 times 3 are how many times 2?

7. Bought 5 cords of pine wood, at $4 a cord, and paid for it with hay, at $10 a ton; how many tons did it take?

8. 5 times 4 are how many times 10?

9. Bought 5 barrels of flour, at $8 a barrel, and paid for it with cloth, at $4 a yard; how many yards did it take?

10. 5 times 8 are how many times 4?

11. How much wool, at 4 shillings a pound, will pay for 8 bushels of corn, at 6 shillings a bushel?

12. 8 times 6 are how many times 4?

13. 7 times 6 are how many times 3?

14. 6 times 6 are how many times 4?

15. How many dozen of eggs, at 12 cents a dozen, will pay for 7 pounds of veal, at 6 cents a pound?

16. 7 times 6 are how many times 12?

17. 8 times 3 are how many times 12? 4? 5? 6?

18. 5 times 7 are how many times 4? 6? 8? 9?

19. 8 times 5 are how many times 10? 7? 6? 9?

20. Bought 3 hundred weight and 3 eighths of a hundred weight of sugar, at 8 dollars a hundred weight, and paid for it with apples, at 3 dollars a barrel; how many barrels did it take? How much broadcloth, at 4 dollars a yard, would it take to pay for it?

21. 3 times 8 and 3 eighths of 8 are how many times 3? 4? 6? 5? 8? 9?

22. Sold 4 bushels and 3 eighths of a bushel of wheat, at 8 shillings a bushel, and took my pay in butter, at 2 shillings a pound; how many pounds did I receive?

23. 4 times 8 and 3 eighths of 8 are how many times 2? 6? 4? 7?

24. How much flour, at $8 a barrel, will pay for 5 boxes and 5 ninths of a box of butter, at $9 a box?

25. 5 times 9 and 5 ninths of 9 are how many times 8? 10? 6? 7?

26. 7 times 10 and 9 tenths of 10 are how many times 8? 11? 6? 12?

27. 5 times 7 and 4 sevenths of 7 are how many times 12? 8? 6? 5? 9?

28. 9 times 8 and 3 eighths of 8 are how many times 5? 10? 12? 7? 11? 3?

29. How many dollars will pay for 8 bushels and 4 sevenths of a bushel of wheat, at 7 shillings a bushel?

30. 8 times 7 and 4 sevenths of 7 are how many times 6? 12? 8? 10? 9?

31. How many bushels of corn, at one dollar a bushel, will pay for 9 bushels and 3 fifths of a bushel of potatoes, at 5 shillings a bushel?

32. 9 times 5 and 3 fifths of 5 are how many times 6? 8? 10? 12? 11?

LESSON III.

1. If 2 pears cost 6 cents, what will 4 pears cost?

Solution. If 2 pears cost 6 cents, 1 pear will cost one half of 6 cents, which is 3 cents; and 4 pears will cost 4 times 3 cents, which are 12 cents; therefore, etc.

Or, since 2 pears cost 6 cents, 2 times 2 pears, that is, 4 pears, will cost 2 times 6 cents, which are 12 cents; therefore, etc.

NOTE. The first mode of answering the above example is the more obvious analysis; but the ready pupil will very often save labor, as in the second solution, by a quick perception of, and a careful attention to, the relations of numbers.

2. 4 times one half of 6 are how many?

3. 6 times one half of 4 are how many?

4. If 3 apples cost 6 cents, what will 12 apples cost?

5. 12 times one third of 6 are how many?

6. 6 times one third of 12 are how many?

7. A man hired a laborer, and agreed to give him $5 for every 2 days' work; how much did he give him for 6 days' work?

8. 5 times one half of 6 are how many?

9. A man lays up 3 shillings a day; in how many days will he lay up $8?

10. 8 times one third of 6 are how many?

11. If 4 boxes of raisins cost $7, what will 12 boxes cost?

12. 7 times one fourth of 12 are how many?

13. If 3 boxes of raisins cost $7, how many boxes may be bought for $35?

14. 3 times one seventh of 35 are how many?

15. If 7 bushels of wheat are worth as much as 2 cords of wood, how many bushels of wheat will pay for 10 cords of wood?

16. 7 times one half of 10 are how many?

17. If 10 cords of wood are worth as much as 35 bushels of wheat, how many bushels of wheat will pay for 2 cords of wood?

18. 35 is how many times one half of 10?

19. How many bushels of corn, at 4 shillings a bushel, will pay for 2 barrels of apples, at 3 dollars a barrel?

20. 2 times 3 times 6 are how many times 4?

21. At 5 shillings a gallon, how many gallons of molasses will pay for 2 yards of cloth, at 5 dollars a yard?

22. 2 times 5 times 6 are how many times 5? 10? 12? 8?

23. How many bushels of wheat, at 8 shillings a bushel, are worth as much as 4 barrels of potatoes, at 2 dollars a barrel?

24. 4 times 2 times 6 are how many times 8? 12? 9? 11?

25. A can build 12 rods of wall in 4 days, and B can build 5 rods as often as A builds 3 rods; how many rods can B build in 4 days?

26. 5 times one third of 12 are how many?

27. A can lay 10 rods of wall in 6 days, and B can lay 7 rods as often as A can lay 5 rods; how many rods can B lay in 9 days?

28. 7 times one fifth of 10 are 2 thirds of what number?

29. A can mow 10 acres of grass in 6 days, but B can mow only 4 acres while A is mowing 5; how many acres can B mow in 3 days?

30. 4 times one fifth of ten are 2 times what number?

31. A boy bought 20 peaches, at the rate of 2 for 3 cents; how many cents did he pay for them?

Solution. Two peaches in 20 peaches 10 times; and 10 times 3 cents are 30 cents; therefore, etc.

Or, 20 peaches, at 3 cents *apiece*, cost 60 cents; but since *two* peaches were bought for 3 cents, the 20 peaches cost *one half* of 60 cents, which is 30 cents.

32. 3 times 20 are how many times 2?

33. How many eggs, at the rate of 3 for 5 cents, can you buy for 30 cents? How many at the rate of 5 for 3 cents?

34. 3 times one fifth of 30 are how many?

35. 3 times 30 are how many times 5?

36. 5 times one third of 30 are how many?

37. 5 times 30 are how many times 3?

38. A hare has 18 rods the start of a hound, but the hound runs 5 rods while the hare runs 3; how many rods must the hound run to overtake the hare?

Solution. The hound gains 2 rods by running 5 rods; he must gain 9 times 2 rods, and therefore must run 9 times 5 rods, which are 45 rods.

39. 5 times one half of 18 are how many?

40. A hare has 24 rods the start of a hound, but the

hound runs 10 rods while the hare runs 7; how many rods will the hare run before the hound overtakes him?

LESSON IV.

1. A BOY having 12 peaches, kept 1 third of them himself, and divided the other 2 thirds equally among 4 of his companions; how many did he give them apiece?

2. Two thirds of 12 are how many times 4?

3. A boy having 21 apples, kept 2 sevenths of them himself, and divided the other 5 sevenths equally among 6 of his playmates; how many did he keep for himself? How many did he give to each of his 6 playmates?

4. Two sevenths of 21 are how many?

5. Five sevenths of 21 are how many times 6?

6. A man gave 4 fifths of 30 cents for 8 oranges; what was the price of 1 orange?

7. Four fifths of 30 are how many times 8?

8. A man who had $36, divided 3 fourths of his money equally among 6 poor persons; how many dollars did he give to each?

9. Three fourths of 36 are how many times 6?

10. A man bought a sheep for 15 shillings, and sold it for 5 thirds of what he gave for it; for how many dollars did he sell it?

11. Five thirds of 15 are how many times 6?

12. A boy bought a rabbit for 25 cents, and sold him for 6 fifths of his cost; for how many dimes did he sell him? How many cents did the boy gain? What part of a dime?

13. Six fifths of 25 are how many times 10?

14. A grocer having 40 pounds of raisins, sold 5 eighths of them in 10 equal parcels; how many pounds were there in each parcel?

15. Five eighths of 40 are how many times 10 ?

16. A boy having 48 cents, paid away 5 eighths of them for oranges at 3 cents apiece ; how many oranges did he buy ? How many cents had he left ?

17. Five eighths of 48 are how many times 3 ?

18. Nine sevenths of 28 are how many times 5 ? 6 ? 8 ?

19. Seven fourths of 36 are how many times 12 ? 6 ? 8 ?

20. Five ninths of 72 are how many times 10 ? 6 ? 7 ? 11 ?

21. Eleven eighths of 64 are how many times 12 ? 10 ? 7 ?

22. Three sevenths of 84 are how many times 9 ? 6 ? 8 ?

23. Five ninths of 72 are how many times 7 ? 6 ? 10 ?

24. Nine fifths of 40 are how many times 12 ? 10 ? 11 ?

25. Nine sevenths of 63 are how many times 10 ? 8 ? 7 ?

LESSON V.

1. CHARLES and Edward bought some oranges: Charles bought 4, which was twice as many as Edward bought; how many did Edward buy ?

2. Four is 2 times what number ?

3. If 2 thirds of a barrel of flour cost 6 dollars, what will 1 third of a barrel cost ?

4. Six is 2 times what number ?

5. If 3 fourths of a ton of coal cost 6 dollars, what will 1 fourth of a ton cost ?

6. Six is 3 times what number ?

7. If 2 fifths of a pound of butter cost 8 cents, what will 1 fifth of a pound cost ?

8. Eight is 2 times what number ?

9. If 3 fifths of a pound of figs cost 12 cents, what will 1 fifth of a pound cost?

10. Twelve is 3 times what number?

11. If 4 fifths of a melon cost 20 cents, what will 1 fifth of it cost?

12. Twenty is 4 times what number?

13. If 5 sixths of a yard of calico cost 15 cents, what will 1 sixth of a yard cost?

14. Fifteen is 5 times what number?

15. If it costs 40 cents to lay 4 sevenths of a rod of wall, what will it cost to lay 1 seventh?

16. Forty is 4 times what number?

17. If 7 eighths of a ton of hay are worth $14, what is 1 eighth of a ton worth?

18. Fourteen is 7 times what number?

19. If 3 fourths of a barrel of flour are worth $6, what is 1 fourth of a barrel worth? If 1 fourth of a barrel is worth $2, what is a barrel worth?

20. If 6 is 3 fourths of some number, what is 1 fourth of the same number? Two is 1 fourth of what number? Then 6 is 3 fourths of what number?

21. If I pay $16 for 4 fifths of a ton of hay, what should I pay for 1 fifth of a ton? If I pay $4 for 1 fifth of a ton, what is the price per ton?

22. If 16 is 4 fifths of some number, what is 1 fifth of the same number? Four is 1 fifth of what number? Then 16 is 4 fifths of what number?

23. If 7 eighths of an acre of land are worth $21, what is 1 eighth of an acre worth? If 1 eighth of an acre is worth $3, what is an acre worth?

24. If 21 is 7 eighths of some number, what is 1 eighth of the same number? Three is 1 eighth of what number? Then 21 is 7 eighths of what number?

25. If 3 fourths of a pound of raisins cost 15 cents, what will 1 fourth of a pound cost? What will a pound cost?

26. Fifteen is 3 fourths of what number?

27. If 5 sevenths of a barrel of fish cost 10 dollars, what will 1 seventh of a barrel cost? What will a barrel cost?

28. Ten is 5 sevenths of what number?

29. If 5 sixths of a barrel of apples cost 20 shillings, how many shillings will a barrel cost?

30. Twenty is 5 sixths of what number?

31. Eighteen is 6 sevenths of what number?

32. Thirty-five is 5 eighths of what number?

33. Twelve is 3 fifths of what number?

34. Fourteen is 7 ninths of what number?

35. Twenty-four is 6 sevenths of what number?

36. Sixteen is 2 fifths of what number?

37. Twenty is 5 ninths of what number?

38. Twenty-seven is 9 tenths of what number?

39. Thirty-two is 8 ninths of what number?

40. Fifteen is 3 sevenths of what number?

41. Thirty-six is 6 sevenths of what number?

42. Forty is 10 elevenths of what number?

43. Forty-two is 6 sevenths of what number.

44. Forty-eight is 8 ninths of what number?

45. Twenty-five is 5 eighths of what number?

46. Fifty-six is 7 ninths of what number?

LESSON VI.

1. If 3 fourths of a pound of raisins cost 15 cents, what will a pound cost? How many peaches, at 2 cents apiece, will it take to pay for a pound of raisins?

2. Fifteen is 3 fourths of how many times 2?

Solution. Since 15 is 3 fourths, 5 is 1 fourth, and 4 fourths will be 4 times 5, which are 20; 2 in 20, 10 times; therefore, 15 is 3 fourths of 10 times 2.

3. If 5 eighths of a ton of hay cost 10 dollars, what will a ton cost? How much cloth, at 4 dollars a yard, will it take to pay for a ton of hay?

4. Ten is 5 eighths of how many times 4?

5. A lady bought 7 eighths of a yard of silk for 21 shillings; how many dollars would a yard cost, at the same rate?

6. Twenty-one is 7 eighths of how many times 6?

7. A man sold a cow for $32, which was only 8 ninths of what she cost him; what did she cost him? When he bought her, he paid for her with wood at $6 a cord; how many cords did it take?

8. Thirty-two is 8 ninths of how many times 6?

9. Samuel bought 18 peaches, which was only 3 eighths as many as Henry bought; how many did Henry buy? Henry paid for his peaches with melons, giving 1 melon for 12 peaches; how many melons did it take?

10. Eighteen is 3 eighths of how many times 12?

11. A man being asked the age of his youngest son, replied that the age of his oldest son was 18 years, which was just 3 eighths of his own age, and that his own age was 16 times the age of his youngest son; what was the father's age? What was the age of his youngest son?

12. Eighteen is 3 eighths of how many times 16?

13. Twenty-four is 3 fifths of how many times 4?

14. Thirty is 5 sixths of how many times 5?

15. Thirty-two is 8 elevenths of how many times 6?

16. Thirty-five is 7 ninths of how many times 10?

17. Thirty-six is 4 sevenths of how many times 8?

18. Forty is 5 sevenths of how many times 9?

19. Forty-two is 6 sevenths of how many times 4?

20. Forty-five is 9 twelfths of how many times 6?

21. Forty-eight is 6 ninths of how many times 12?

22. Fifty is 5 sixths of how many times 8?

23. Forty-nine is 7 eighths of how many times 10?

24. Sixty is 5 sevenths of how many times 8?

25. Sixty-three is 7 eighths of how many times 12?

26. Sixty-four is 8 ninths of how many times 6?

27. Sixty-five is 5 sixths of how many times 10?

28. Seventy is 10 sevenths of how many times 8?
29. Seventy-two is 12 fifths of how many times 4?
30. Seventy-five is 5 fourths of how many times 9?
31. Eighty is 10 sevenths of how many times 6?
32. Eighty is 8 thirds of how many times 8?

LESSON VII

1. A BOY gave away 2 apples, which was one third of all he had; how many had he at first? How many did he keep?

2. Sarah is 5 years old, and she is one fourth as old as her brother; how old is her brother? How much older is he than Sarah?

3. A boy gave away 10 cents, which was 2 thirds of all the money he had; how many cents had he?

4. A boy sold a dove for 15 cents, which was 3 fourths of what it cost him; how much did he lose by the bargains?

5. If 15 is 3 fourths of some number, what is 1 fourth of the same number?

6. A man paid away $12, which was 3 fifths of all the money he had; how many dollars did he keep?

7. If 12 is 3 fifths of some number, what is 2 fifths of the same number?

8. A man sold a watch for $24, which was only 3 fifths of what it cost him; what did it cost him?

9. Twenty-four is 3 fifths of what number?

10. A man sold a horse for $30, which was only 5 sevenths of what he paid for him; how much did he lose by his bargains?

11. If 30 is 5 sevenths of some number, what is 2 sevenths of the same nnmber? Thirty is 5 sevenths of what number?

12. A man sold 3 watches for $36, which was 6 fifths of what they cost him; what did they cost him apiece?

13. Thirty-six is 6 fifths of how many times 3?

14. There is a pole standing so that 1 third of it is under water, and 8 feet above water; what part of the pole is above water? How long is the pole?

15. There is a pole standing so that 2 thirds of it are under water, and 5 feet out; how long is the pole?

16. There is a pole 3 fifths under water, and 8 feet out; what part of the pole is above water? How long is the pole?

17. There is a pole standing so that 2 sevenths of it are in the mud, 3 sevenths in the water, and 6 feet above water? What part of the pole is above water? How long is the pole? How many feet are in the mud? How many in the water?

18. There is an orchard in which 5 ninths of the trees bear apples, 2 ninths bear peaches, 1 ninth bear pears, and 6 trees bear cherries; how many trees are there in the orchard? How many of each kind?

19. Three sevenths of the trees in a certain orchard bear cherries, 2 sevenths bear peaches, 10 trees bear pears, and 6 bear plums; how many trees are there in the orchard? How many of each kind?

20. In a certain school 3 ninths of the pupils study arithmetic, 2 ninths study grammar, 1 ninth study history, 1 ninth study algebra, and 8 pupils study geometry; how many pupils are there in the school? How many attending to each study?

21. There is a school in which 3 elevenths of the scholars study Latin, 2 elevenths study Greek, 4 elevenths study arithmetic, 9 scholars study algebra, and 7 study geometry; how many scholars are there in the school? How many attending to each study?

22. A man sold a cow for $35, which was 7 fifths of what she cost him; how much did he gain by the bargains?

23. If 35 is 7 fifths of some number, what is 2 fifths of the same number?

LESSON VIII.

1. Two boys were counting their money, when one said he had 6 cents. Well, said the other, 2 thirds of your money is exactly 1 fifth of mine; now, if you will tell me how many cents I have, I will give you one half of them. How many cents had he?

2. Two thirds of 6 are 1 fifth of what number?

3. A boy being asked how many chickens he had, said that he had them in two coops: in one coop he had 12, and 3 fourths of these were just 1 fifth of what he had in the other; required the number in the other coop?

4. Three fourths of 12 are 1 fifth of what number?

5. Two boys were talking of their ages, when one of them said he was twelve years old. Well, said the other, 2 thirds of your age are exactly 4 fifths of mine; and if you will tell me how old I am, I will give you as many apples as I am years old. What was his age?

6. Two thirds of 12 are 4 fifths of what number?

7. A man being asked the age of his oldest son, replied that his youngest son was 8 years old, and that 3 fourths of the youngest son's age were just 2 sevenths of the age of his oldest son; how old was the oldest son?

8. Three fourths of 8 are 2 sevenths of what number?

9. A man being asked how many sheep he had, said that he had them in two pastures; in one he had 32, and 5 eighths of these were 2 ninths of what he had in the other pasture; how many had he in the other?

10. 5 eighths of 32 are 2 ninths of what number?

11. 3 fifths of 10 are 2 sevenths of what number?

12. 4 sevenths of 21 are 3 eighths of what number?

13. 5 sixths of 24 are 4 ninths of what number?

14. 5 thirds of 24 are 4 sevenths of what number?

15. 8 sevenths of 21 are 4 fifths of what number?

16. Charles is 35 months old, and 4 fifths of his age are 7 sixths of Mary's age; how many years old is Mary?

17. 4 fifths of 35 are 7 sixths of how many times 12?

18. 4 fifths of 25 are 10 sevenths of how many times 5?

Solution. 1 fifth of 25 is 5, and 4 fifths are 4 times 5, which are 20; if 20 is 10 sevenths, 1 seventh is 1 tenth of 20, which is 2, and 7 sevenths are 7 times 2, which are 14; 5 in 14, 2 times and 4 fifths: therefore 4 fifths of 25 are 10 sevenths of 2 times 5 and 4 fifths of 5.

19. 4 sevenths of 28 are 2 fifths of how many times 9?

20. 3 fifths of 30 are 9 tenths of how many times 8?

21. 3 eighths of 32 are 6 sevenths of how many times 9?

22. 6 ninths of 36 are 8 fifths of how many times 4?

23. 3 fifths of 40 are 6 elevenths of how many times 7?

24. 8 ninths of 45 are 10 sixths of how many times 3?

25. 5 eighths of 48 are 3 fourths of how many times 12?

26. 3 eighths of 64 are 4 ninths of how many times 10?

27. 5 ninths of 72 are 4 fifths of how many times 6?

LESSON IX.

1. Four sevenths of 42 are 8 ninths of how many fifths of 35?

Solution. 1 seventh of 42 is 6, and 4 sevenths are 4 times 6, which are 24; if 24 is 8 ninths, 1 ninth is 1 eighth of 24, which is 3, and 9 ninths are 9 times 3, which are 27; 1 fifth of 35 is 7, and 7 in 27, 3 times and 6 sevenths of a time; therefore, 4 sevenths of 42 are 8 ninths of 3 fifths and 6 sevenths of a fifth of 35.

2. 3 fifths of 20 are 6 sevenths of how many eighths of 24 ?

3. 3 sevenths of 28 are 4 fifths of how many ninths of 18 ?

4. 3 eighths of 32 are 6 tenths of how many sixths of 42 ?

5. 4 ninths of 81 are 6 sevenths of how many thirds of 24 ?

6. 2 eighths of 72 are 6 fifths of how many ninths of 36 ?

7. 5 eighths of 56 are 5 ninths of how many tenths of 100 ?

8. 3 eighths of 64 are 6 sevenths of how many sevenths of 35 ?

9. 9 sevenths of 28 are 3 fifths of how many sixths of 48 ?

10. 3 tenths of 100 are 5 ninths of how many fourths of 32 ?

11. 5 eighths of 32 are 10 sixths of how many fifths of 20 ?

12. 3 fourths of 48 are 6 sevenths of how many thirds of 24 ?

LESSON X.

1. If 1 barrel of flour costs $6, what will 4 barrels cost ?

2. If 2 barrels of apples are worth $6, what are 8 barrels worth ?

3. What will 3 yards of cloth cost, if 8 yards cost $32 ?

4. Two couriers, A and B, are 21 miles apart, and traveling in the same direction. A, who is before B, travels 5 miles, and B travels 8 miles an hour ; in how many hours will B overtake A ? How many miles will A go before B overtakes him ? How far must B go to overtake A ?

5. Two couriers, 12 miles apart, are traveling in the same direction; but the foremost goes only 3 miles, while the other goes 5 miles; how far will each go before they will be together?

6. A and B are 60 miles apart, and traveling towards each other; while A goes 4 miles, B goes 6; how many miles will each travel before they meet?

7. A boy having 25 apples, kept 4 himself, and divided the rest equally among 3 companions; how many did he give them apiece?

8. How many chairs, at $3 each, may be bought for $36?

9. How many sheep, at $8 each, may be bought for $48?

10. How many tables, at $9 each, may be bought for $63?

11. A man bought 3 barrels of flour, at $7 a barrel, 5 yards of cloth, at $4 a yard, and 1 ton of coal for $9; what did he pay for all?

12. If 4 tons of hay will keep 2 cows through the winter, how many tons will keep 12 cows the same time?

13. If a man spends 8 shillings a day, how many dollars will he spend in 6 days? In 9 days?

14. If a man earns $60 in 12 weeks, how much does he earn in 1 month, or 4 weeks? How much in 3 weeks? In 5 weeks?

15. Bought 5 pieces of cloth, each containing 10 yards, for $100; what was the cost of each piece? What the price per yard?

16. If 2 men spend $24 in 3 weeks, how many dollars, at the same rate, will 5 men spend in 4 weeks?

17. If 4 horses eat 12 bushels of oats in 3 weeks, how many horses will eat 24 bushels in the same time?

18. If 2 horses eat 4 bushels of oats in 8 days, how many bushels will 3 horses eat in 12 days?

19. If 6 horses eat 9 bushels of oats in 5 days, in how many days will 3 horses eat 18 bushels?

20. A ship's crew of 9 men have provisions for 2 months; how many months will the same provisions last 6 men?

21. A ship has food for 2 men for 12 months; how many men will it last 4 months?

22. If a staff 3 feet long casts a shadow 4 feet at 9 o'clock, what is the length of a pole that casts a shadow 20 feet, at the same time?

23. If a staff 3 feet long casts a shadow 2 feet at 12 o'clock, what is the length of the shadow cast, at the same time, by a pole 18 feet long?

24. If 4 men can do a piece of work in 6 days, in how many days can 3 men do the same?

25. If 5 men can do a piece of work in 3 days, in how many days can they do a piece of work 5 times as large?

26. If 36 men can do a piece of work in 4 days, how many men can do twice as much work in 2 days?

27. A man bought 3 barrels of flour for 21 dollars; at what price per barrel must he sell it to gain 6 dollars on the lot?

28. A man bought 5 barrels of flour, at $6 a barrel; for what sum must the whole lot be sold to gain $10?

29. A man bought 5 casks of nails, at 7 dollars a cask; at what price per cask must he sell them to gain 15 dollars?

30. A man bought 6 yards of cloth for 24 dollars, and sold it at 6 dollars a yard; how much did he gain by the bargains?

31. A boy bought 8 oranges, at 3 cents apiece, and sold them all for 40 cents; how much did he gain by the bargains?

32. A man bought a lot of hay, at $12 a ton, and

sold it at $14 a ton, by which he gained $16; how many tons did he buy?

33. By buying cloth at $4 a yard, and selling it at $3 a yard, I lost $6; what did I pay for all of the cloth?

34. A cabinet-maker exchanged 3 bureaus, worth $8 apiece, for chairs worth $2 apiece; how many chairs did he receive?

35. B can build 5 rods of wall while A builds 4; how many rods can B build while A is building 24 rods?

LESSON XI.

1. At 7 dollars a barrel, what will 3 barrels and 1 half of a barrel of flour cost?

2. 3 times 7 and 1 half of 7 are how many?

3. If coal is worth $9 a ton, what shall I pay for 4 tons and 1 fourth of a ton? What for 5 tons and 3 fourths of a ton?

4. 5 times 9 and 3 fourths of 9 are how many?

5. Bought 5 yards and 3 eighths of a yard of velvet, at $8 a yard, and paid for it with broadcloth, at $4 a yard; how many yards of broadcloth did it take?

6. 5 times 8 and 3 eighths of 8 are how many times 4?

7. Bought 3 tons and 5 eighths of a ton of hay, at $16 a ton, and paid for it with flour, at $8 a barrel; how many barrels did it take?

8. 3 times 16 and 5 eighths of 16 are how many times 8?

9. 5 times 12 and 5 sixths of 12 are how many times 8? 9? 6? 5?

10. 7 times 10 and 3 fifths of 10 are how many times 12? 9? 11? 8?

11. 6 times 9 and 2 thirds of 9 are how many times 6? 10? 8? 12? 7?

12. If 6 barrels of apples cost $18, what part of $18 will 1 barrel cost? What part of $18 will 2 barrels cost? 3 barrels? 4 barrels? 5 barrels?

13. What is 1 sixth of 18? 1 third of 18? 1 half of 18? 2 thirds of 18? 5 sixths of 18?

14. If 8 yards of cloth cost $24, what part of $24 will 1 yard cost? 2 yards? 3 yards? 4 yards? 5 yards? 6 yards? 7 yards?

15. What is 1 eighth of 24? 1 fourth of 24? 3 eighths of 24? 1 half of 24? 5 eighths of 24? 3 fourths of 24? 7 eighths of 24?

16. If 12 cords of wood cost $48, what part of $48 will 1 cord cost? 2 cords? 3 cords? 4 cords? 5 cords? 6 cords? 7 cords? 8 cords? 9 cords? 10 cords? 11 cords?

17. What is 1 twelfth of 48? 1 sixth of 48? 1 fourth of 48? 1 third of 48? 5 twelfths of 48? 1 half of 48? 7 twelfths of 48? 2 thirds of 48? 3 fourths of 48? 5 sixths of 48? 11 twelfths of 48?

18. A boy having 28 cents, gave 5 sevenths of them for peaches, at 2 cents apiece; how many peaches did he buy?

19. 5 sevenths of 28 are how many times 2? 5?

20. A man bought a sheep for 21 shillings, and sold it for 8 sevenths of what he paid for it; how many dollars did he receive for the sheep?

21. 8 sevenths of 21 are how many times 6? 3? 12? 8? 4?

22. Six men, setting out on a journey, took 5 loaves of bread apiece; but before they had eaten any of it 4 other men joined them, and the bread was divided equally among the whole company. How many loaves did each man have?

23. Six times 5 are how many times 10? 8? 4? 3?

24. A man had 24 cords of wood, and sold 3 eighths of it for $54; what was the price per cord?

25. 54 are how many times 3 eighths of 24?

26. A man had 36 sheep, and sold 5 ninths of them at 4 dollars a head; how many dollars did he receive?

27. How many are 4 times 5 ninths of 36?

28. If 5 eighths of a box of raisins cost 15 shillings, how many bushels of oats, at 4 shillings a bushel, will pay for a box of raisins?

29. 8 fifths of fifteen are how many times 4? 6?

30. A man gave 25 cents for his breakfast, which was 5 eighths of what he gave for his dinner; what did he give for his dinner?

31. A man paid 45 cents for his dinner, which was 9 sevenths of what he paid for his breakfast; what did his breakfast cost?

32. A teacher, when asked how many pupils he had, replied that they were in 2 rooms: in one room there were 32, and 5 eighths of these were 2 fifths of the number in the other; how many pupils were there in the other room? How many in both?

33. 5 eighths of 32 are 2 fifths of what number?

34. A boy being asked how old his dog was, said that his cat was 8 years old, that the age of his cat was 4 thirds of the age of his rabbit, and that his dog was twice as old as his rabbit; how old was his dog?

LESSON XII.

1. How many are 2 times 13? 13 times 2?
2. How many are 2 times 14? 14 times 2?
3. How many are 2 times 15? 15 times 2?
4. How many are 2 times 16? 16 times 2?
5. How many are 2 times 17? 17 times 2?
6. How many are 2 times 18? 18 times 2?
7. How many are 2 times 19? 19 times 2?
8. How many are 2 times 20? 20 times 2?
9. How many are 2 times 21? 21 times 2?
10. How many are 2 times 22? 22 times 2?
11. How many are 2 times 23? 23 times 2?

12. How many are 2 times 24? 24 times 2?
13. How many are 2 times 25? 25 times 2?
14. How many are 3 times 13? 13 times 3?
15. How many are 3 times 14? 14 times 3?
16. How many are 3 times 15? 15 times 3?
17. How many are 3 times 16? 16 times 3?
18. How many are 3 times 17? 17 times 3?
19. How many are 3 times 18? 18 times 3?
20. How many are 3 times 19? 19 times 3?
21. How many are 3 times 20? 20 times 3?
22. How many are 3 times 21? 21 times 3?
23. How many are 3 times 22? 22 times 3?
24. How many are 3 times 23? 23 times 3?
25. How many are 3 times 24? 24 times 3?
26. How many are 3 times 25? 25 times 3?
27. How many are 4 times 13? 13 times 4?
28. How many are 4 times 14? 14 times 4?
29. How many are 4 times 15? 15 times 4?
30. How many are 4 times 16? 16 times 4?
31. How many are 4 times 17? 17 times 4?
32. How many are 4 times 18? 18 times 4?
33. How many are 4 times 19? 19 times 4?
34. How many are 4 times 20? 20 times 4?
35. How many are 4 times 21? 21 times 4?
36. How many are 4 times 22? 22 times 4?
37. How many are 4 times 23? 23 times 4?
38. How many are 4 times 24? 24 times 4?
39. How many are 4 times 25? 25 times 4?
40. How many are 5 times 13? 13 times 5?
41. How many are 5 times 14? 14 times 5?
42. How many are 5 times 15? 15 times 5?
43. How many are 5 times 16? 16 times 5?
44. How many are 5 times 17? 17 times 5?
45. How many are 5 times 18? 18 times 5?
46. How many are 5 times 19? 19 times 5?
47. How many are 5 times 20? 20 times 5?
48. How many are 5 times 21? 21 times 5?

49. How many are 5 times 22? 22 times 5?
50. How many are 5 times 23? 23 times 5?
51. How many are 5 times 24? 24 times 5?
52. How many are 5 times 25? 25 times 5?
53. How many are 6 times 13? 13 times 6?
54. How many are 6 times 14? 14 times 6?
55. How many are 6 times 15? 15 times 6?
56. How many are 6 times 16? 16 times 6?
57. How many are 6 times 17? 17 times 6?
58. How many are 6 times 18? 18 times 6?
59. How many are 6 times 19? 19 times 6?
60. How many are 6 times 20? 20 times 6?
61. How many are 6 times 21? 21 times 6?
62. How many are 6 times 22? 22 times 6?
63. How many are 6 times 23? 23 times 6?
64. How many are 6 times 24? 24 times 6?
65. How many are 6 times 25? 25 times 6?
66. Count by 6's to 96; thus, 6, 12, 18, etc.
67. Count backward by 6's; thus, 96, 90, 84, etc.
68. Count by 6's from 3 to 99; thus, 3, 9, 15, etc.
69. Count backward by 6's; thus, 99, 93, 87, etc.
70. Count by 7's to 98; thus, 7, 14, 21, etc.
71. Count backward by 7's; thus, 98, 91, 84, etc.

NOTE. Such exercises as the last six examples above should be continued only a very few minutes at a time. Let the exercise be varied by allowing the class to recite sometimes in concert, sometimes individually; by allowing one pupil to name one number, and another the next; by letting two or more pupils recite simultaneously, one counting from one number, and another from another number; or by any other mode which shall secure the attention and awaken the interest of the class.

The more complicated of the above plans, if adopted, should be commenced with the 2's, the 5's, the 10's, and other numbers which give the more simple combinations; and they should not be continued unless the pupils have the ability to proceed without confusing each other.

These exercises, if judiciously conducted, are very valuable; and when the pupil shall have acquired the ability to make all such combinations with accuracy and rapidity, he will have very great facility in all the processes of addition and subtraction.

SECTION SEVENTH.

LESSON I.

1. A MERCHANT sold 5 yards of cloth, at $\frac{1}{3}$ of a dollar per yard; what did he receive for it?

2. What is 5 times $\frac{1}{3}$? *Ans.* $\frac{1}{3} \times 5 = \frac{5}{3} = 1\frac{2}{3}$.

3. How many are $\frac{2}{7} \times 3$? $\frac{3}{7} \times 5$?

4. How many are $\frac{2}{9} \times 4$? $\frac{5}{9} \times 7$?

5. How many are 3 times $\frac{2}{9}$?

Ans. Three times $\frac{2}{9}$ are $\frac{6}{9}$, or $\frac{2}{3}$.

NOTE. The $\frac{6}{9}$ are obtained by multiplying the numerator by 3, and the $\frac{2}{3}$ are obtained by *dividing the denominator* by 3. In examples like this, it is better to divide the denominator, *because it gives the answer in smaller or lower terms.*

6. How many are 5 times $\frac{2}{15}$? Why?

7. At $\frac{4}{15}$ of a dollar a pound, what will 5 pounds of butter cost?

Solution. Five pounds will cost 5 times as much as 1 pound; 5 times $\frac{4}{15}$ of a dollar are $\frac{4}{3}$ of a dollar $= \$1\frac{1}{3}$,

8. At $\frac{3}{8}$ of a dollar per pound, what will 4 pounds of butter cost?

9. What will 6 yards of silk cost, at $\frac{1}{12}$ of a dollar per yard?

10. If 1 horse eats $\frac{8}{21}$ of a ton of hay in a month, how much will 7 horses eat in the same time?

11. If 1 man can reap $\frac{19}{24}$ of an acre of rye in a day, how much can 8 men reap in the same time?

12. If 1 yard of cloth costs $\frac{15}{16}$ of a dollar, what will 8 yards cost?

13. How many are 8 times $\frac{15}{16}$? 9 times $\frac{17}{18}$?

14. How many are 6 times $\frac{23}{24}$? 11 times $\frac{19}{22}$?

15. How many are 5 times $\frac{3}{5}$?

Ans. Five times $\frac{3}{5}$ are $\frac{3}{1}$, or 3.

NOTE. The pupil will observe that *if a fraction is multiplied by its denominator, the product is the numerator.*

16. How many are 11 times $\frac{9}{11}$? Why?

17. How many are 9 times $\frac{5}{9}$? 15 times $\frac{7}{15}$?

18. How many are 23 times $\frac{17}{23}$? 49 times $\frac{11}{49}$?

19. What cost 8 boxes of strawberries, at $\frac{3}{8}$ of a dollar per box?

20. What cost 25 yards of sheeting, at $\frac{7}{25}$ of a dollar per yard?

21. If it costs $\frac{13}{20}$ of a dollar to build a rod of wall, what will it cost to build 20 rods?

22. If a boy runs $\frac{14}{15}$ of a rod in 1 second, how many rods, at the same rate, will he run in 15 seconds?

23. If a locomotive runs $\frac{32}{75}$ of a mile in 1 minute, how far will it run in 75 minutes?

24. If 2 bushels of apples cost $\frac{6}{7}$ of a dollar, what will 1 bushel cost?

Solution. One bushel will cost 1 half as much as 2 bushels; 1 half of $\frac{6}{7}$ of a dollar is $\frac{3}{7}$ of a dollar; therefore, if 2 bushels cost $\frac{6}{7}$ of a dollar, 1 bushel will cost $\frac{3}{7}$ of a dollar.

NOTE. To find $\frac{1}{2}$ of a number is the same as to divide the number by 2. It is just as evident that $\frac{1}{2}$ of $\frac{6}{7}$ is $\frac{3}{7}$, as it is that $\frac{1}{2}$ of 6 cents is 3 cents. Hence, *dividing the numerator of a fraction by any number, is dividing the fraction by the same number.*

25. If a boy walks 3 miles in $\frac{6}{7}$ of an hour, how long will it take him to walk 1 mile?

26. What is $\frac{1}{3}$ of $\frac{6}{7}$? *Ans.* $\frac{6}{7} \div 3 = \frac{2}{7}$.

27. If 4 men can reap $\frac{8}{15}$ of an acre of rye in an hour, how much can 1 man reap in the same time?

28. If a boy can run $\frac{12}{25}$ of a mile in 3 minutes, how far can he run in 1 minute?

29. What is 1 dozen of eggs worth, if 7 dozen are worth $\frac{14}{15}$ of a dollar?

30. What is $\frac{1}{7}$ of $\frac{14}{15}$? *Ans.* $\frac{14}{15} \div 7 = \frac{2}{15}$.

31. What is $\frac{6}{11} \div 2$? $\frac{12}{17} \div 4$? $\frac{20}{23} \div 5$?

32. What is $\frac{15}{4} \div 5$? $\frac{24}{7} \div 6$? $\frac{40}{3} \div 8$?

33. Hannah divided one third of a pie equally between 2 children; what part of the whole pie did she give to each child?

Solution. One third equals $\frac{2}{6}$, and $\frac{1}{2}$ of $\frac{2}{6}$ is $\frac{1}{6}$; that is, $\frac{1}{2}$ of $\frac{1}{3}$ is $\frac{1}{6}$; and therefore she gave $\frac{1}{6}$ of the whole pie to each.

NOTE. Multiplying the denominator by 2 makes twice as many parts in the unit, and therefore each part is only one half as great; hence, *multiplying the denominator by any number, divides the fraction by the same number.*

34. Three fifths of a barrel of flour were divided equally among 4 poor families; what part of a barrel did each receive?

35. If $\frac{7}{8}$ of a bushel of chestnuts are divided equally among 5 boys, what part of a bushel does each receive?

36. What is 1 sixth of $\frac{5}{9}$? 1 eighth of $\frac{5}{7}$?

37. What is 1 ninth of $\frac{5}{8}$? 1 twelfth of $\frac{7}{11}$?

38. Mary having a nice large pineapple, gave $\frac{1}{11}$ of it to Sarah, $\frac{2}{11}$ to Fanny, $\frac{4}{11}$ to Laura, and kept the rest herself; what part of the pineapple did she give away? What part did she keep?

39. What is $\frac{1}{11}+\frac{2}{11}+\frac{4}{11}$?

40. What is $\frac{11}{11}-\frac{7}{11}$? $\frac{11}{11}-\frac{4}{11}$? $\frac{7}{11}+\frac{4}{11}$?

41. What is $\frac{1}{15}+\frac{2}{15}+\frac{4}{15}+\frac{7}{15}$? $\frac{15}{15}-\frac{14}{15}$?

42. What is $\frac{8}{23}+\frac{9}{23}+\frac{3}{23}-\frac{6}{23}$? $\frac{23}{23}-\frac{14}{23}$?

LESSON II.

1. ONE half is equal to how many sixths?

Solution. Since $\frac{6}{6}$ make a whole one, $\frac{1}{2}$ of a whole one will be $\frac{1}{2}$ of $\frac{6}{6}$, which is $\frac{3}{6}$; therefore, etc.

2. One third of an apple is equal to how many sixths of an apple?

3. James gave $\frac{1}{2}$ of a pear to William, $\frac{1}{3}$ to George, and kept the rest himself; how much did he give away? How much did he keep?

NOTE. Change the fraction to *sixths*. Before fractions that have *different* denominators can be added together, or before one can be subtracted from another, it is necessary to change them to other fractions that have *like* denominators.

4. $\frac{1}{2}+\frac{1}{3}=$ what? $1-\frac{5}{6}=$ what?

5. A man gave $\frac{1}{2}$ of a dollar to one of his workmen, and $\frac{3}{4}$ of a dollar to another; how many fourths of a dollar did he give to both? How many dollars?

6. $\frac{1}{2}+\frac{3}{4}$ are how many $\frac{1}{4}$? How many times 1?

NOTE. Read the first question in Ex. 6 as follows: One half and three fourths are how many *fourths?* Read all similar examples in like manner.

7. A man gave $\frac{1}{2}$ of a barrel of apples to one of his neighbors, and $\frac{3}{6}$ of a barrel to another; did he give more to one than to the other?

8. $\frac{1}{2}$ and $\frac{3}{6}$ are how many $\frac{1}{6}$? How many times 1?

9. Henry gave $\frac{1}{3}$ of an orange to his brother John, $\frac{2}{6}$ to his sister Marion, and kept the rest himself. With this division the selfish John was dissatisfied, saying that Henry had given more to Marion than to him. No, John, said Marion, Henry has given just as much to you as to me. Now which was right, John or Marion? What part of the orange did Henry keep for himself?

10. $\frac{1}{3}$ is how many $\frac{1}{6}$?

11. $\frac{1}{3}+\frac{2}{6}+\frac{1}{3}=$ how many $\frac{1}{6}$? How many times 1?

12. Samuel worked $\frac{2}{3}$ of a day for Mr. Adams, and $\frac{4}{6}$ of a day for Mr. Daniels. Mr. Adams paid Samuel 50 cents. Now, at the same rate, what should Mr. Daniels pay him? How much, at that rate, will Samuel earn in a day? What part of a dollar?

13. $\frac{2}{3}$ are how many $\frac{1}{6}$?

14. A boy wished to give $\frac{1}{2}$ of a pear to his sister, and $\frac{1}{3}$ to his brother; and in order to do this most conveniently, he first cut the pear into 6 equal pieces; how many pieces did he give to each? How many pieces had he remaining?

15. Samuel gave $\frac{1}{3}$ of a dollar to one poor boy, and $\frac{1}{4}$ of a dollar to another; what part of a dollar did he give away? How much more to one boy than to the other?

16. What is $\frac{1}{3}+\frac{1}{4}$? $\frac{1}{3}-\frac{1}{4}$?

17. James had a box of strawberries, and said he would give $\frac{1}{4}$ of them to Addie, $\frac{1}{3}$ to Willie, and the rest to Georgie, if he could tell how to divide them. Georgie won; how did he divide them?

18. Harriet having 3 quarts of cherries, said she would give $\frac{5}{8}$ of a quart to Mary, $\frac{5}{4}$ of a quart to Martha, and the rest to any one of the company who would most quickly tell how to divide them. Would you have won had you been there, and how would you divide them?

19. $\frac{5}{4}$ are how many $\frac{1}{8}$? 3 are how many $\frac{1}{8}$?

20. $\frac{5}{4}+\frac{5}{8}=$ how many $\frac{1}{8}$? $3-\frac{15}{8}=$ what?

21. A man having 2 bushels of wheat to give to 3 of his workmen, wished to give $\frac{2}{3}$ of a bushel to the first, $\frac{3}{5}$ of a bushel to the second, and the rest to the third; how much should he give to the third?

22. $\frac{2}{3}$ are how many $\frac{1}{15}$? $\frac{3}{5}$ are how many $\frac{1}{15}$?

23. A farmer had a horse, a cow, and a sheep. The horse would eat $\frac{3}{8}$ of a ton of hay in a month, the cow $\frac{3}{10}$, and the sheep $\frac{3}{40}$; how much would they all eat? How much more would the horse eat than the sheep? How much less would the sheep eat than the cow?

24. $\frac{3}{8}$ are how many $\frac{1}{40}$? $\frac{3}{10}$ are how many $\frac{1}{40}$?

25. A man gave $\frac{5}{8}$ of a ton of coal to B, and $\frac{3}{4}$ of a ton to C; to which did he give the most? How much? How much to both?

26. A little girl having a pound of figs, gave $\frac{4}{9}$ of them to one schoolmate, $\frac{5}{12}$ to another, and kept the rest; what part of the pound did she give away? What part did she keep?

27. $\frac{4}{9}$ are how many $\frac{1}{36}$? $\frac{5}{12}$ are how many $\frac{1}{36}$?

28. A spendthrift having received a fortune, squandered $\frac{5}{12}$ of it the first month, and $\frac{7}{20}$ of it the next month, after receiving it; what part of it had he then remaining?

LESSON III.

1. REDUCE $\frac{2}{4}$ to its lowest terms. *Ans.* $\frac{1}{2}$.
2. Reduce $\frac{4}{6}$ to its lowest terms.
3. Reduce $\frac{6}{9}$ to its lowest terms.
4. Reduce $\frac{6}{8}$ to its lowest terms.
5. Reduce $\frac{6}{12}$ to its lowest terms.
6. Reduce $\frac{9}{12}$ to its lowest terms.
7. Reduce $\frac{10}{15}$ to its lowest terms.
8. Reduce $\frac{9}{15}$ to its lowest terms.
9. Reduce $\frac{12}{15}$ to its lowest terms
10. Reduce $\frac{12}{18}$ to its lowest terms.
11. Reduce $\frac{9}{21}$ to its lowest terms.
12. Reduce $\frac{24}{32}$ to its lowest terms.
13. Reduce $\frac{14}{21}$ to its lowest terms.
14. Reduce $\frac{14}{35}$ to its lowest terms.
15. Reduce $\frac{18}{27}$ to its lowest terms.
16. Reduce $\frac{33}{44}$ to its lowest terms.
17. Reduce $\frac{36}{45}$ to its lowest terms.
18. Reduce $\frac{27}{63}$ to its lowest terms.
19. Reduce $\frac{1}{3}$ to twelfths. $\frac{1}{4}$ to twelfths.
20. Reduce $\frac{2}{3}$ to twelfths. $\frac{3}{4}$ to twelfths.
21. Reduce $\frac{6}{10}$ to fifths. $\frac{3}{5}$ to twentieths.
22. Reduce $\frac{9}{12}$ to fourths. $\frac{3}{4}$ to twentieths.
23. Reduce $\frac{3}{5}$ and $\frac{3}{4}$ to equivalent fractions having a common denominator.
24. Reduce $\frac{6}{10}$ and $\frac{9}{12}$ so that they shall have the least common denominator.
25. Reduce $\frac{2}{3}$ and $\frac{3}{5}$. $\frac{8}{12}$ and $\frac{12}{20}$.
26. Reduce $\frac{2}{7}$ and $\frac{5}{6}$. $\frac{6}{21}$ and $\frac{10}{12}$.
27. Reduce $\frac{3}{7}$ and $\frac{4}{5}$. $\frac{12}{28}$ and $\frac{12}{15}$.
28. Reduce $\frac{1}{2}$, $\frac{1}{3}$, and $\frac{1}{4}$, so that they shall have the least common denominator.
29. Reduce $\frac{3}{6}$, $\frac{5}{15}$, and $\frac{6}{24}$.
30. Reduce $\frac{3}{8}$, $\frac{5}{6}$, and $\frac{1}{4}$.
31. Add together $\frac{1}{2}$ and $\frac{2}{3}$. $\frac{3}{4}$ and $\frac{2}{5}$.
32. Add together $\frac{3}{7}$ and $\frac{2}{3}$. $\frac{4}{5}$ and $\frac{1}{3}$.

33. Add together $\frac{5}{8}$ and $\frac{5}{6}$. $\frac{10}{16}$ and $\frac{15}{18}$.

34. Add together $\frac{1}{2}$, $\frac{2}{5}$, and $\frac{7}{10}$. $\frac{2}{4}$, $\frac{6}{15}$, and $\frac{28}{40}$.

35. Subtract $\frac{1}{3}$ from $\frac{1}{2}$. $\frac{1}{2}$ from $\frac{2}{3}$.

36. Subtract $\frac{3}{10}$ from $\frac{4}{5}$. $\frac{4}{5}$ from $\frac{9}{10}$.

37. Subtract $\frac{1}{7}$ from $\frac{2}{3}$. $\frac{5}{8}$ from $\frac{9}{10}$.

38. Subtract $\frac{3}{5}$ from $\frac{6}{7}$. $\frac{5}{9}$ from $\frac{11}{12}$.

39. A and B engage to do a piece of work. A can do $\frac{1}{2}$ of it in a day, and B can do $\frac{1}{4}$ of it; what part of the work can they together do in a day? What part of it would remain for the next day? What part of the next day would it take them to finish the work?

40. A, B, and C engage to do a piece of work. A can do $\frac{1}{3}$ of it in a day, B can do $\frac{2}{5}$ of it, and C can do $\frac{1}{6}$ of it; what part of the work can they all do in a day? What part would remain for the next day? What part of the second day would it take them to finish the work?

41. $\frac{1}{10}$ is what part of $\frac{9}{10}$? $\frac{3}{30}$ what part of $\frac{27}{30}$?

LESSON IV.

1. A BOY having $\frac{1}{2}$ of an apple, gave away $\frac{1}{2}$ of that; what part of the whole apple did he give away?

2. What is $\frac{1}{2}$ of $\frac{1}{2}$? $\frac{1}{2}$ is how many $\frac{1}{4}$?

3. A boy having $\frac{1}{2}$ of a melon, gave away $\frac{1}{3}$ of that; what part of the whole melon did he give away? What part did he keep?

4. What is $\frac{1}{3}$ of $\frac{1}{2}$? $\frac{1}{2}$ — $\frac{1}{3}$ of $\frac{1}{2}$?

5. James having $\frac{1}{3}$ of a day for play, spent $\frac{1}{2}$ of that time with his cousin Alfred; what part of the whole day was he with Alfred?

6. What is $\frac{1}{2}$ of $\frac{1}{3}$? $\frac{1}{3}$ of $\frac{1}{2}$?

7. A man having $\frac{1}{2}$ of a cord of wood, gave away $\frac{1}{4}$ of that; what part of a cord did he give away? What part did he keep?

8. What is $\frac{1}{4}$ of $\frac{1}{2}$? $\frac{3}{4}$ of $\frac{1}{2}$?

Ans. $\frac{1}{4}$ of $\frac{1}{2}$ is $\frac{1}{8}$; and $\frac{3}{4}$ of $\frac{1}{2}$ is 3 times $\frac{1}{8}$, or $\frac{3}{8}$.

9. A man having $\frac{1}{4}$ of a cord of wood, gave away $\frac{1}{2}$ of that, and sold the rest; what part of a cord did he give away? What part did he sell?

10. What is $\frac{1}{2}$ of $\frac{1}{4}$?

11. Which is the most, $\frac{1}{4}$ of $\frac{1}{2}$, or $\frac{1}{2}$ of $\frac{1}{4}$?

12. What is $\frac{1}{4}$ of $\frac{1}{3}$? $\frac{1}{3}$ of $\frac{1}{4}$?

13. A merchant owning $\frac{1}{4}$ of a ship, sold $\frac{1}{4}$ of his share; what part of the ship did he sell? What part did he keep?

14. What is $\frac{1}{4}$ of $\frac{1}{4}$? $\frac{3}{4}$ of $\frac{1}{4}$?

15. What is $\frac{1}{4}$ of $\frac{1}{5}$? $\frac{3}{4}$ of $\frac{1}{5}$?

16. What is $\frac{1}{5}$ of $\frac{1}{7}$? $\frac{4}{5}$ of $\frac{1}{7}$?

17. What is $\frac{1}{6}$ of $\frac{1}{9}$? $\frac{5}{6}$ of $\frac{1}{9}$?

18. A boy having $\frac{3}{4}$ of a melon, gave away $\frac{1}{2}$ of what he had; what part of a whole melon did he give away?

Solution. If $\frac{1}{4}$ of a melon is cut into 2 equal parts, each part is $\frac{1}{8}$ of the whole melon; and if $\frac{1}{2}$ of $\frac{1}{4}$ is $\frac{1}{8}$, then $\frac{1}{2}$ of $\frac{3}{4}$ is $\frac{3}{8}$; therefore, he gave away $\frac{3}{8}$ of the whole melon.

19. A man owning $\frac{5}{8}$ of a ship, sold $\frac{1}{3}$ of his share; what part of the whole ship did he sell?

20. A girl having picked $\frac{5}{8}$ of a pailful of blueberries, spilled $\frac{1}{4}$ of them; what part of a pailful did she spill?

21. A boy having $\frac{3}{4}$ of a dollar, lost $\frac{1}{5}$ of what he had; what part of a dollar did he lose?

22. What is $\frac{1}{5}$ of $\frac{3}{4}$? $\frac{1}{3}$ of $\frac{5}{8}$? $\frac{1}{7}$ of $\frac{3}{10}$?

23. What is $\frac{1}{8}$ of $\frac{5}{9}$? $\frac{1}{9}$ of $\frac{4}{7}$? $\frac{1}{10}$ of $\frac{7}{9}$?

24. What is $\frac{1}{6}$ of $\frac{7}{11}$? $\frac{1}{7}$ of $\frac{5}{7}$? $\frac{1}{12}$ of $\frac{5}{7}$?

25. A boy having $\frac{4}{5}$ of a pear, gave away $\frac{2}{3}$ of what he had; what part of a whole pear did he give away?

Solution. $\frac{1}{3}$ of $\frac{1}{5}$ is $\frac{1}{15}$; therefore, $\frac{1}{3}$ of $\frac{4}{5}$ is $\frac{4}{15}$; and if $\frac{1}{3}$ of $\frac{4}{5}$ is $\frac{4}{15}$, then $\frac{2}{3}$ of $\frac{4}{5}$ is 2 times $\frac{4}{15}$, which are $\frac{8}{15}$; therefore, he gave away $\frac{8}{15}$ of the whole pear.

26. Charles had $\frac{3}{4}$ of a dollar, and gave $\frac{3}{5}$ of that for a knife; what part of a dollar did the knife cost?

27. If a yard of lace is worth $\frac{7}{8}$ of a dollar, what are $\frac{3}{4}$ of a yard worth?

28. What cost $\frac{3}{8}$ of a bushel of corn, at $\frac{7}{8}$ of a dollar per bushel?

29. A man owning $\frac{7}{8}$ of a ship, sold $\frac{3}{5}$ of his share; what part of the whole ship did he sell? What part did he keep?

30. If a boy can pick $\frac{5}{8}$ of a bushel of chestnuts in a day, what part of a bushel can he pick in $\frac{3}{4}$ of a day?

31. If a man can reap $\frac{8}{9}$ of an acre of wheat in a day, what part of an acre can he reap in $\frac{5}{8}$ of a day?

Solution. In $\frac{1}{8}$ of a day he can reap $\frac{1}{8}$ of $\frac{8}{9}$, which is $\frac{1}{9}$ of an acre; and in $\frac{5}{8}$ of a day he can reap 5 times $\frac{1}{9}$, which are $\frac{5}{9}$, of an acre.

32. What cost $\frac{3}{5}$ of a bushel of butternuts, at $\frac{5}{7}$ of a dollar per bushel?

33. What cost $\frac{7}{8}$ of a bushel of corn, at $\frac{24}{25}$ of a dollar per bushel?

34. If a man can mow $\frac{10}{33}$ of an acre of grass in an hour, what part of an acre can he mow in $\frac{3}{5}$ of an hour?

35. If a dozen of oranges are worth $\frac{12}{17}$ of a dollar, what are $\frac{5}{6}$ of a dozen worth?

36. At $\frac{15}{16}$ of a dollar per day, what will a man earn in $\frac{3}{5}$ of a day?

37. If a boy learns $\frac{8}{17}$ of his lesson in an hour, what part of it does he learn in $\frac{3}{4}$ of an hour?

38. At $\frac{14}{25}$ of a dollar per rod, what will it cost to build $\frac{6}{7}$ of a rod of fence?

39. A man owning $\frac{2}{3}$ of a ship, sold $\frac{1}{2}$ of his share; how many thirds of the ship did he sell?

40. What is $\frac{1}{2}$ of $\frac{2}{3}$?

41. A man owning $\frac{4}{5}$ of a farm, sold $\frac{1}{4}$ of his share; what part of the farm did he sell? What part did he keep?

42. What is $\frac{1}{4}$ of $\frac{4}{5}$? $\frac{3}{4}$ of $\frac{4}{5}$?

43. What is $\frac{1}{6}$ of $\frac{6}{7}$? $\frac{5}{6}$ of $\frac{6}{7}$?

44. What is $\frac{1}{3}$ of $\frac{6}{7}$? $\frac{2}{3}$ of $\frac{6}{7}$?

45. What is $\frac{1}{4}$ of $\frac{12}{13}$? $\frac{3}{4}$ of $\frac{12}{13}$?

46. What is $\frac{1}{5}$ of $\frac{10}{21}$? $\frac{4}{5}$ of $\frac{10}{21}$?

47. What is $\frac{1}{2}$ of $\frac{2}{3}$ of $\frac{3}{4}$?

Solution. $\frac{1}{2}$ of $\frac{2}{3}$ is $\frac{1}{3}$; $\frac{1}{3}$ of $\frac{3}{4}$ is $\frac{1}{4}$; therefore, $\frac{1}{2}$ of $\frac{2}{3}$ of $\frac{3}{4}$ is $\frac{1}{4}$.

48. What is $\frac{1}{2}$ of $\frac{2}{3}$ of $\frac{3}{4}$ of $\frac{4}{5}$ of $\frac{5}{6}$ of $\frac{6}{7}$ of $\frac{7}{8}$ of $\frac{8}{9}$?

49. A girl wishing to divide $\frac{4}{5}$ of a pineapple equally among her 3 brothers, was at a loss to know how to divide it. At length she decided to cut each of the fifths into 3 equal parts; how many pieces did she give each of the boys? What part of the pineapple was each piece?

50. The girl mentioned in the above example could have made a better division of the $\frac{4}{5}$ of a pineapple; can you tell how? By your plan, how many pieces would you give to each boy? What would be the size of the pieces?

LESSON V.

1. IF a barrel of apples cost $2\frac{1}{4}$ dollars, what will $\frac{1}{2}$ of a barrel cost?

2. What is $\frac{1}{2}$ of $2\frac{1}{4}$?

Solution. $\frac{1}{2}$ of 2 is 1; $\frac{1}{2}$ of $\frac{1}{4}$ is $\frac{1}{8}$; therefore, $\frac{1}{2}$ of $2\frac{1}{4}$ is $1\frac{1}{8}$.

3. If 3 barrels of flour cost $\$21\frac{3}{4}$, what is the price per barrel?

4. What is $\frac{1}{3}$ of $21\frac{3}{4}$?

5. A lady divided $8\frac{3}{5}$ pounds of maple sugar equally among her 4 children; how much did she give to each?

6. What is $\frac{1}{4}$ of $8\frac{3}{5}$?

7. A boy divided $2\frac{1}{2}$ oranges equally among his brother, his sister, and himself; what part of an orange did each have?

8. What is $\frac{1}{3}$ of $2\frac{1}{2}$?

Solution. $2\frac{1}{2}$ are the same as $\frac{5}{2}$, and $\frac{1}{3}$ of $\frac{5}{2}$ is $\frac{5}{6}$; therefore, $\frac{1}{3}$ of $2\frac{1}{2}$ is $\frac{5}{6}$.

9. If 4 gallons of molasses cost $\$3\frac{1}{3}$, what does 1 gallon cost?

10. What is $\frac{1}{4}$ of $3\frac{1}{3}$?

11. If 6 bushels of oats cost $\$3\frac{3}{5}$, what is the price per bushel?

12. What is $\frac{1}{6}$ of $3\frac{3}{5}$?

13. Paid $\$7\frac{3}{4}$ for 10 yards of cloth; what did I pay for 1 yard?

14. What is $\frac{1}{10}$ of $7\frac{3}{4}$?

15. Bought 12 bushels of corn for $\$11\frac{4}{5}$; what did it cost per bushel?

16. What is $\frac{1}{12}$ of $11\frac{4}{5}$?

17. If 5 bushels of wheat cost $\$8\frac{3}{4}$, what is the price per bushel?

18. What is $\frac{1}{5}$ of $8\frac{3}{4}$?

Solution. $8\frac{3}{4}$ equals $\frac{35}{4}$, and $\frac{1}{5}$ of $\frac{35}{4}$ is $\frac{7}{4}$, or $1\frac{3}{4}$.

Or, 5 in $8\frac{3}{4}$, once and $3\frac{3}{4}$ over; $3\frac{3}{4}$ equals $\frac{15}{4}$, and $\frac{1}{5}$ of $\frac{15}{4}$ is $\frac{3}{4}$, which, added to the 1, gives $1\frac{3}{4}$.

19. A man bought 6 pounds of cheese for $62\frac{1}{2}$ cents, what did he pay per pound?

20. What is $\frac{1}{6}$ of $62\frac{1}{2}$? $\frac{1}{6}$ of $74\frac{2}{3}$?

21. If 7 barrels of flour cost $60\frac{3}{4}$ dollars, what is the price per barrel? What is the cost of 2 barrels?

22. What is $\frac{1}{7}$ of $60\frac{3}{4}$? $\frac{2}{7}$ of $60\frac{3}{4}$?

23. A man bought a cow for $\$33\frac{1}{3}$, and afterward sold her for $\frac{5}{4}$ of her cost; how many dollars did he gain? What did he receive for her?

24. What is $\frac{1}{4}$ of $33\frac{1}{3}$? $\frac{5}{4}$ of $33\frac{1}{3}$?

25. A man bought a horse for $\$65\frac{3}{5}$; but, not proving so good as he anticipated, he sold it for $\frac{7}{8}$ of its cost; how much did he lose? For how many dollars did he sell it?

26. What is $\frac{1}{8}$ of $65\frac{3}{5}$? $\frac{7}{8}$ of $65\frac{3}{5}$?

27. If 1 man can do a piece of work in $27\frac{3}{4}$ days, how long would it take 4 men to do the same?

28. What is $\frac{1}{4}$ of $27\frac{3}{4}$? $\frac{1}{2}$ of $\frac{1}{4}$ of $27\frac{3}{4}$?

29. If a horse can travel $27\frac{3}{5}$ miles in 4 hours, how far, at the same rate, can he travel in 9 hours?

30. What is $\frac{1}{4}$ of $27\frac{3}{5}$? $\frac{9}{4}$ of $27\frac{3}{5}$?

31. If 7 yards of cloth cost $\$31\frac{2}{3}$, what will 5 yards cost?

32. What is $\frac{1}{7}$ of $31\frac{2}{3}$? $\frac{5}{7}$ of $31\frac{2}{3}$?

33. If 4 pipes will empty a cistern containing $38\frac{2}{5}$ gallons in a given time, how many gallons will be drawn from the cistern by 3 pipes of the same size in the same time?

34. What is $\frac{1}{4}$ of $38\frac{2}{5}$? $\frac{3}{4}$ of $38\frac{2}{5}$?

35. If 3 horses eat $22\frac{1}{2}$ bushels of oats in 8 weeks, what will 7 horses eat in the same time?

36. What is $\frac{7}{3}$ of $22\frac{1}{2}$? $\frac{3}{5}$ of $18\frac{2}{3}$?

37. In a freshet, a tree stood so that $35\frac{2}{3}$ feet, which were $\frac{3}{7}$ of its entire length, were under water; how many feet were above water? What was the whole length of the tree?

38. What is $\frac{4}{3}$ of $35\frac{2}{3}$? $\frac{7}{3}$ of $35\frac{2}{3}$?

39. A man distributed $23\frac{4}{5}$ bushels of wheat equally among 7 poor families; how many bushels, at the same rate, would have been required for 9 families?

40. What is $\frac{9}{7}$ of $23\frac{4}{5}$? $\frac{7}{9}$ of $13\frac{4}{5}$?

41. If 8 soldiers are allowed $13\frac{1}{3}$ pounds of meat in 1 day, what will 3 soldiers be allowed in 5 days?

42. What is 5 times $\frac{3}{8}$ of $13\frac{1}{3}$?

43. What is 4 times $\frac{2}{7}$ of $19\frac{1}{2}$?

LESSON VI.

1. A BOY having 2 melons, ate $\frac{1}{3}$ of a melon each day as long as they lasted; how many days did they last? How many days would they have lasted had he eaten $\frac{2}{3}$ of a melon each day?

2. How many times is $\frac{1}{3}$ contained in 2; that is how many times is $\frac{1}{3}$ contained in $\frac{6}{3}$?

Solution. 6 times; for 2 equals $\frac{6}{3}$, and $\frac{1}{3}$ is contained in $\frac{6}{3}$ just as many times as 1 apple is contained in 6 apples, or as 1 is contained in 6.

3. How many times $\frac{2}{3}$ in 2?

Solution. 3 times; for $\frac{2}{3}$ is contained in $\frac{6}{3}$ just as many times as 2 in 6. Or, since $\frac{1}{3}$ is contained 6 times in $\frac{6}{3}$, $\frac{2}{3}$ is contained $\frac{1}{2}$ of 6 times, viz., 3 times, in $\frac{6}{3}$.

4. A boy having 3 oranges, gave $\frac{3}{4}$ of an orange apiece to several of his companions; to how many could he give them?

5. How many times $\frac{1}{4}$ in 3? $\frac{3}{4}$ in 3?

6. If $\frac{2}{3}$ of a pound of butter will last a family 1 day, how many days will 4 pounds last the same family? How many days will 6 pounds last? 7 pounds?

7. How many times $\frac{2}{3}$ in 4? In 6? In 7?

8. If $\frac{2}{5}$ of a barrel of apples will supply a family for a week, for how many weeks will 6 barrels supply the same family? For how many weeks will 6 barrels supply them, if they use $\frac{3}{5}$ of a barrel a week? If they use $\frac{4}{5}$?

9. In 6 how many times $\frac{2}{5}$? $\frac{3}{5}$? $\frac{4}{5}$?

10. If a family consumes $\frac{3}{4}$ of a barrel of flour in a month, how long will $3\frac{3}{4}$ barrels last them? $9\frac{3}{4}$ barrels? $8\frac{1}{4}$?

11. How many times $\frac{3}{4}$ in $3\frac{3}{4}$? In $9\frac{3}{4}$? $8\frac{1}{4}$?

12. How long will 9 pounds of butter last a family, if they use $1\frac{1}{2}$ pounds per week?

13. How many times $1\frac{1}{2}$, or $\frac{3}{2}$ in 9?

14. How many soldiers will $8\frac{3}{4}$ pounds of bread supply, if each soldier is allowed $1\frac{3}{4}$ pounds?

15. In $8\frac{3}{4}$ how many times $1\frac{3}{4}$?

16. How many garments will $20\frac{1}{4}$ yards of cloth make, if each garment requires $2\frac{1}{4}$ yards?

17. In $20\frac{1}{4}$ how many times $2\frac{1}{4}$?

18. When raisins cost $\frac{1}{6}$ of a dollar a pound, how many pounds can you buy for $\frac{3}{4}$ of a dollar?

19 How many times is $\frac{1}{6}$ contained in $\frac{3}{4}$?

Solution. $\frac{3}{4}$ are the same as $\frac{9}{12}$, and $\frac{1}{6}$ equals $\frac{2}{12}$; $\frac{2}{12}$ in $\frac{9}{12}$, $4\frac{1}{2}$ times; therefore, $\frac{1}{6}$ in $\frac{3}{4}$, $4\frac{1}{2}$ times.

20. When cream of tartar costs $\frac{3}{5}$ of a dollar a pound, how many pounds may be bought for $\$1\frac{1}{2}$? How many for $\$2\frac{3}{4}$?

21. How many times $\frac{3}{5}$ in $1\frac{1}{2}$? In $2\frac{3}{4}$?

22. At $\frac{5}{8}$ of a dollar per pound, how many pounds of tea can I buy for $\$2\frac{1}{2}$? For $\$3\frac{1}{4}$?

23. How many times $\frac{5}{8}$ in $2\frac{1}{2}$? In $3\frac{1}{4}$?

24. If coffee costs $\frac{3}{8}$ of a dollar a pound, how many pounds may be bought for $\$1\frac{5}{6}$? For $\$2\frac{2}{3}$?

25. How many times $\frac{3}{8}$ in $1\frac{5}{6}$? In $2\frac{2}{3}$?

26. At $\$1\frac{1}{2}$ per bushel, how much wheat can I buy for $\$3\frac{1}{4}$? For $\$5\frac{1}{8}$?

27. How many times $1\frac{1}{2}$ in $3\frac{1}{4}$? In $5\frac{1}{8}$?

1st *Ans.* $1\frac{1}{2} = \frac{3}{2} = \frac{6}{4}$; $3\frac{1}{4} = \frac{13}{4}$; $\frac{13}{4} \div \frac{6}{4} = \frac{13}{6} = 2\frac{1}{6}$.

28. At $\$2\frac{1}{4}$ a box, how many boxes of raisins can you buy for $\$6\frac{3}{4}$? For $\$9\frac{3}{8}$?

29. How many times $2\frac{1}{4}$ in $6\frac{3}{4}$? In $9\frac{3}{8}$?

30. At $8\frac{1}{3}$ cents a pound, how many pounds of sugar can I buy for $12\frac{1}{2}$ cents? For $33\frac{1}{3}$ cents?

31. How many times $8\frac{1}{3}$ in $12\frac{1}{2}$? In $33\frac{1}{3}$?

32. How many times $6\frac{1}{4}$ in $37\frac{1}{2}$? In $87\frac{1}{2}$?

33. How many times $5\frac{3}{8}$ in $6\frac{5}{12}$? In $7\frac{1}{3}$?

34. If $3\frac{2}{5}$ barrels of flour cost $\$20\frac{2}{5}$, what will $6\frac{4}{5}$ barrels cost?

Suggestion. $6\frac{4}{5}$ is twice $3\frac{2}{5}$.

35. If $5\frac{3}{4}$ barrels of apples cost $\$11\frac{1}{2}$, what will $28\frac{3}{4}$ barrels cost?

36. If $2\frac{3}{5}$ acres of land are worth $26, what are $10\frac{2}{5}$ acres worth?

37. If $7\frac{2}{9}$ acres of land cost $\$28\frac{8}{9}$, what will $5\frac{3}{4}$ acres cost?

Suggestion. $7\frac{2}{9}$ in $28\frac{8}{9}$, 4 times.

38. If $2\frac{3}{4}$ pounds of veal cost $24\frac{3}{4}$ cents, what will $5\frac{1}{3}$ pounds cost? $10\frac{2}{3}$ pounds? $21\frac{1}{3}$ pounds? $7\frac{2}{3}$ pounds?

SECTION EIGHTH.

LESSON I.

REMARKS. 1. *Per centum*, or *per cent.*, is an expression which means *by the hundred;* thus, 5 per cent. of any number is $\frac{5}{100}$ or $\frac{1}{20}$ of that number; 25 per cent. is $\frac{25}{100}$ or $\frac{1}{4}$; 3 per cent. is $\frac{3}{100}$, etc.

2. Gains and losses, interest on money, commissions of agents, increase or decrease of population, etc., are usually computed at some given per cent. on the sum or number considered.

3. The following rates, being readily reduced to convenient fractional forms, are easily computed:

2 per cent. is	$\frac{1}{50}$.	$12\frac{1}{2}$ per cent. is	$\frac{1}{8}$.
4 "	$\frac{1}{25}$.	$16\frac{2}{3}$ "	$\frac{1}{6}$.
5 "	$\frac{1}{20}$.	20 "	$\frac{1}{5}$.
$6\frac{1}{4}$ "	$\frac{1}{16}$.	25 "	$\frac{1}{4}$.
$8\frac{1}{3}$ "	$\frac{1}{12}$.	$33\frac{1}{3}$ "	$\frac{1}{3}$.
10 "	$\frac{1}{10}$.	50 "	$\frac{1}{2}$.

Ex. 1. A farmer having 80 sheep, lost 25 per cent. of them; how many sheep did he lose?

Solution. He lost $\frac{25}{100}$ or $\frac{1}{4}$ of his flock; $\frac{1}{4}$ of 80 is 20; therefore, he lost 20 sheep.

2. The number of inhabitants in the town of B on the 1st of May, 1860, was 2500; in a year the population increased 20 per cent.; what was the increase of population in the year? What was the population of B on the 1st of May, 1861?

3. An agent sells $500 worth of goods for his employer, and receives a commission, or compensation, of 4 per cent. for transacting the business; how many dollars does he receive for his services? How many dollars does he pay over to his employer?

4. In one town in Wisconsin 16000 bushels of wheat grew in 1863; what will be the number of bushels in 1864, if there shall be a decrease of $12\frac{1}{2}$ per cent.?

5. A physician having bills against his patients to the amount of $72, paid $16\frac{2}{3}$ per cent. for their collection; what did the agent receive for collecting the money? What did he pay over to the physician?

6. Suppose the number of deaths in New York during the 1st week of March was 500, and that the decrease during the 1st week of June was 10 per cent.; how many less deaths were there in the city during the latter week? What was the number of deaths during the latter week?

7. The American Insurance Company has insured $4000 on my house, at a premium of 2 per cent.; what is the premium, or sum paid for the insurance?

8. What is $6\frac{1}{4}$ per cent. of $64?

9. What is $33\frac{1}{3}$ per cent. of 45 bushels of corn?

10. What is 5 per cent. of 80 peaches?

11. What is $8\frac{1}{3}$ per cent. of 48 boys?

12. What is 50 per cent. of 36 horses?

13. A physician having bills against his patients to the amount of $80, paid $20 for collecting them; what per cent. did he pay?

Solution. $80 is 100 per cent., or the whole sum collected, and $20 is $\frac{20}{80}$ or $\frac{1}{4}$ of the whole sum; $\frac{1}{4}$ of 100 per cent. is 25 per cent.; therefore, etc.

14. In a lot of 60 barrels of apples, 12 barrels are spoiled and the rest are good; what per cent. of them are spoiled? What per cent. are good?

15. From a cask of 63 gallons of molasses, a grocer has drawn 21 gallons; what per cent. of the molasses has he drawn out? What per cent. of it remains in?

16. What per cent. do I gain by buying hats at $4, and selling them at $5, apiece?

17. What per cent. do I lose by buying hats at $5, and selling them at $4, apiece?

18. Twenty is what per cent. of 50?

19. A gentleman invested $\frac{2}{5}$ of his money in land; what per cent. of it did he so invest?

Solution. The whole of any thing is 100 per cent. of it; hence $\frac{2}{5}$ of the thing is $\frac{2}{5}$ of 100 per cent., which is 40 per cent.; therefore, etc.

20. A man saved $\frac{3}{8}$ of his income, and spent the rest; what per cent. of it did he save? What per cent. did he spend?

21. Three fourths of a number is what per cent. of it?

22. Seven eighths of a number is what per cent. of it?

23. Five sixths of a number is what per cent. of it?

24. At 5 per cent., what sum can be insured on a ship bound from Boston to San Francisco, for a premium of $60?

Solution. Since 5 per cent. is $\frac{5}{100}$ or $\frac{1}{20}$ of the sum insured, the premium, $60, is $\frac{1}{20}$ of the sum insured; $60 is $\frac{1}{20}$ of $1200; therefore, etc.

25. At 2 per cent., what sum can be insured on a house for a premium of $50?

26. $24 is $16\frac{2}{3}$ per cent. of what sum?

27. Seven oranges are 25 per cent. of what?

28. In 1863, a farmer raised 63 bushels of corn on an acre. This was an increase of $12\frac{1}{2}$ per cent. on the previous year's crop; how many bushels did he raise on the acre in 1862?

Solution. Since the increase was $12\frac{1}{2}$ per cent., or $\frac{1}{8}$, of the crop, 63 bushels is $\frac{9}{8}$ of the crop; 63 bushels is $\frac{9}{8}$ of 56 bushels; therefore, the crop of 1862 was 56 bushels to the acre.

29. A grocer sold tea at 50 cents per pound, and thereby gained 25 per cent. on the cost; what was the cost per pound?

30. A grocer sold tea at 45 cents per pound, and thereby lost 25 per cent. on the cost; what was the cost per pound?

31. A merchant sends his agent $42, from which he is to take a commission and expend the rest for coffee. The agent's commission being 5 per cent. on the purchase, what sum will he have to expend?

Solution. The sum to be expended is $\frac{100}{100}$ of itself, and the commission is $\frac{5}{100}$ of the same sum; hence the sum sent to the agent is $\frac{105}{100}$ or $\frac{21}{20}$ of the sum expended. \$42 is $\frac{21}{20}$ of \$40; therefore, etc.

32. A man sends his agent in Oregon \$515, from which he is to take a commission and invest the balance in land; the commission being 3 per cent. on the purchase, what sum was invested in land?

33. If $12\frac{1}{2}$ per cent. is gained by selling hay at \$18 per ton, what did it cost?

34. If $16\frac{2}{3}$ per cent. is lost by selling a watch for \$40, what did it cost?

35. An agent retained a commission of $12\frac{1}{2}$ per cent. for collecting money, and paid \$56 over to his employer; what sum did he collect?

Solution. Since $12\frac{1}{2}$ per cent. or $\frac{1}{8}$ of the sum collected was retained by the agent, the \$56 paid over was $\frac{7}{8}$ of the sum collected. \$56 is $\frac{7}{8}$ of \$64; therefore, etc.

36. A farmer engaged a laborer to thresh his wheat, agreeing to give him 10 per cent. of all he threshed; how many bushels must he thresh that the farmer may keep 54 bushels?

37. If it requires \$125 in paper currency to buy \$100 in gold, how much gold may be bought for \$100 in paper currency?

NOTE. When it takes more than \$100 in bank bills, government bonds, or other paper currency, to buy \$100 in gold, gold is said to be at a *premium* if the paper currency is taken as the standard of value, and the paper currency is said to be at a *discount* if gold is taken as the standard.

38. If it takes \$150 in paper currency to buy \$100 in gold, what is the per cent. of discount on the paper?

Solution. The paper is worth $\frac{100}{150} = \frac{2}{3}$ as much as gold; therefore the discount on the paper is $\frac{1}{3} = 33\frac{1}{3}$ per cent. of its nominal value.

39. If it takes $\$166\frac{2}{3}$ of paper to buy \$100 in gold,

at what per cent. of premium is gold, paper being taken as the standard? At what per cent. of discount is paper, gold being taken as the standard?

LESSON II.

REMARKS. 1. INTEREST is money paid *for the use* of borrowed money. The money borrowed is called the *Principal.* The sum of the principal and interest is called the *Amount.*

2. In many of the United States interest is computed at 6 per cent., and this will be the per cent. understood in this book if no other is mentioned.

3. In computing interest, thirty days are usually considered a month, and 12 months a year.

Ex. 1. The interest of $1 for a year being 6 cents, what is the interest of $1 for 2 years? For 3 years? 4 years? 5 years? 8 years? 10 years? 15 years?

2. What is the interest of a dollar for 1 month?

Solution. Since the interest of $1 for 12 months is 6 cents, the interest of $1 for 1 month is $\frac{1}{12}$ of 6 cents, which is $\frac{6}{12}$ or $\frac{1}{2}$ of a cent = 5 mills.

3. What is the interest of $1 for 2 months? For 3 months? 4 months? 6 months? 9 months? 5 months? 8 months? 11 months? 10 months? 7 months?

4. What is the interest of $1 for 1yr. 2m.? For 1yr. 6m.? 1yr. 3m.? 1yr. 8m.?

5. What is the interest of $1 for 2yr. 6m.? 2yr. 3m.? 2yr. 10m.? 3yr. 4m.? 5yr. 7m.?

6. What is the interest of $1 for 6 days?

Solution. Since the interest of $1 for 30 days is 5 mills, the interest of $1 for 6 days is $\frac{6}{30}$ or $\frac{1}{5}$ of 5 mills, which is 1 mill.

7. What is the interest of $1 for 12 days? For 18 days? 24 days? 1m. 6d.? 1m. 24d.? 2m. 18d.? 4m. 12d.? 7m. 18d.? 8m. 24d.?

8. What is the interest of $1 for 1 day?

Solution. Since the interest of $1 for 6 days is 1 mill, the interest of $1 for 1 day is $\frac{1}{6}$ of a mill.

9. What is the interest of $1 for 2 days? For 3 days? 4 days? 5 days? 7 days? 8 days? 9 days? 15 days? 20 days? 28 days? 19 days?

10. What is the interest of $1 for 1m. 15d.? 2m. 3d.? 3m. 9d.? 8m. 4d.? 7m. 27d.? 1yr. 6m. 15d.?

11. What is the interest of $2 for a year?

Solution. Since the interest of $1 for a year is 6 cents, the interest of $2 is 2 times 6 cents, which are 12 cents.

12. What is the interest of $3 for a year? Of $4? Of $5? $10? $12? $20? $30?

13. What is the interest of $2 for 1yr. 6m.? For 1yr. 6m. 24d.? 2yr. 3m. 6d.? 3yr. 5m. 15d.?

14. What is the interest of $3 for 2yr. 6m. 12d.? Of $5 for 2yr. 3m. 15d.? Of $10 for 1yr. 7m. 14d.?

15. What is the interest of $100 for 1yr.? For 2 yr.? 1yr. 6m.? 1yr. 10m.? 2yr. 3m.?

16. What is the interest of $200 for 1yr.? For 2yr.? 1yr. 4m.? 2yr. 6m.? 3yr. 2m.? 4yr. 8m.?

17. What is the interest of $50 for 1yr.? For 2yr.? 5yr.? 6m.? 8m.? 1yr. 4m.?

18. What is the interest of $25 for 1yr.? 3yr.? 1yr. 6m.? 2yr. 8m.?

19. What is the interest of $100 for 60 days?

Solution. The interest of $1 for 60 days is 1 cent; therefore the interest of $100 is 100 cents, or $1.

20. What is the interest of $30 for 60 days? For 30 days? 20 days? 40 days?

21. What is the interest of $40 for 60 days? For 30 days? 15 days? 45 days? 75 days?

22. What is the interest of $20 for 33 days

Suggestion. 33 days $=$ 30 days $+ \frac{1}{10}$ of 30 days.

23. What is the interest of $40 for 33 days? For 66 days? 90 days? 93 days?

24. The interest of any sum of money for 1 year, at 5 per cent., is what part of the principal?

Solution. 5 per cent. is $\frac{5}{100}=\frac{1}{20}$; therefore, the interest of any sum for 1 year at 5 per cent. is $\frac{1}{20}$ of the principal.

25. What is the interest of $40 for 1 year at 5 per cent.?

26. At 4 per cent., the interest of any sum for 1 year equals what part of the principal?

27. At 10 per cent., the interest equals what part of the principal? At 12½ per cent.? At 20 per cent.? At 16⅔ per cent.?

28. What is the interest of $24 for 1 year at 16⅔ per cent.? At 12½ per cent.? At 25 per cent.?

29. What is the interest of $75 for 1 year at 4 per cent.? At 8 per cent.? At 33⅓ per cent.?

30. What is the interest of $60 for 1 year and 6 months at 5 per cent.?

31. What is the interest of $32 for 1 year and 6 months at 12½ per cent.?

32. At 6¼ per cent., what is the interest of $32 for 1 year? For 2 years? 1 year and 6 months?

33. At 8⅓ per cent., what is the interest of $72 for 2 years? For 1 year and 6 months?

LESSON III.

1. WHAT is the *amount* of $50 for 1 year?

Solution. The *interest* of $50 for 1 year is $3, therefore, the *amount* is $50 + $3 = $53.

2. What is the interest of $200 for 2 years and 6 months? What the amount?

3. What is the amount of $60 for 60 days?

4. What is the amount of $80 for 33 days?

5. What is the amount of $84 for 90 days?

6. What is the amount of $50, at 4 per cent., for 2 years and 6 months?

7. What is the amount of $60, at $8\frac{1}{3}$ per cent., for 3 years and 3 months?

8. What is the amount of $64, at $12\frac{1}{2}$ per cent., for 2 years and 9 months?

9. What is the amount of $36, at $16\frac{2}{3}$ per cent., for 2 years and 3 months?

10. What is the amount of $84, at 25 per cent., for 2 years and 6 months?

11. At what *rate per cent.* will $6 gain 45 cents interest in $1\frac{1}{2}$ years?

Solution. The interest of $6 for 1 year, at 1 per cent., is 6 cents, and for $1\frac{1}{2}$ years it is $1\frac{1}{2}$ times 6 cents, or 9 cents. If $6, at 1 per cent., gains 9 cents, then to gain 45 cents will require as many per cent. as 9 cents is contained times in 45 cents: 9 cents in 45 cents, 5 times; therefore, $6, at 5 per cent., will gain 45 cents in $1\frac{1}{2}$ years.

Or, thus: Since the interest for $1\frac{1}{2}$ years or $\frac{3}{2}$ years is 45 cents, for $\frac{1}{2}$ year it is $\frac{1}{3}$ of 45 cents, or 15 cents, and for $\frac{2}{2}$ years or a whole year it is 2 times 15 cents, or 30 cents; 30 cents is $\frac{30}{600}=\frac{5}{100}$ of $6; $\frac{5}{100}$ is 5 per cent.; therefore, $6, at 5 per cent., gains 45 cents in $1\frac{1}{2}$ years.

12. At what rate per cent. will $8 gain 96 cents in 1 year and 6 months; that is, in $1\frac{1}{2}$ years?

13. At what per cent. will $100 gain $$10\frac{1}{2}$ in $1\frac{1}{2}$ years?

14. At what per cent. will $50 gain $10 in 2 years?

15. At what per cent. will $12 gain 80 cents in 1 year and 4 months?

16. At what per cent. will a given principal double itself in 20 years?

Solution. Any principal will double itself in 1 year, at 100 per cent.; and, therefore, in 20 years, at $\frac{1}{20}$ of 100 per cent., viz., 5 per cent.

17. At what per cent. will any principal double itself in 25 years? In 10 years? $12\frac{1}{2}$ years? 50 years? $16\frac{2}{3}$ years? $33\frac{1}{3}$ years? 5 years? 4 years?

18. In what time will \$6, on interest at 5 per cent., gain 70 cents?

Solution. \$6, at 5 per cent., gains 30 cents in 1 year; 30 cents in 70 cents, $2\frac{1}{3}$ times; therefore, \$6, at 5 per cent., will gain 70 cents in $2\frac{1}{3}$ years, or 2 years and 4 months.

19. In what time, at 6 per cent., will \$10 gain \$3?

20. In what time, at 10 per cent., will \$100 gain \$25?

21. In what time, at 4 per cent., will \$4 gain 24 cents?

22. In what time, at 8 per cent., will \$50 gain \$23?

23. In what time, at 5 per cent., will a given principal double itself?

Solution. Any principal, at 1 per cent., will double itself, i. e. it will gain 100 per cent., in 100 years; and at 5 per cent., in $\frac{1}{5}$ of 100 years, viz., 20 years.

24. In what time, at 4 per cent., will any principal double itself? In what time, at 6 per cent.? At 8 per cent.? At 20 per cent.? At 25 per cent.? At 2 per cent.? At 50 per cent.? At $12\frac{1}{2}$ per cent.? At $6\frac{1}{4}$ per cent.? At $16\frac{2}{3}$ per cent.? At $33\frac{1}{3}$ per cent.?

LESSON IV.

1. What principal, at 6 per cent., will gain 36 cents in 1 year and 6 months?

Solution. \$1, at 6 per cent., gains 9 cents in 1 year and 6 months; 9 cents in 36 cents, 4 times; therefore, \$4, at 6 per cent., will gain 36 cents in 1 year and 6 months.

Or, since the interest of any sum for 1 year and 6 months, at 6 per cent., is $\frac{9}{100}$ of the principal, it follows that 36 cents is $\frac{9}{100}$ of the required principal; hence, $\frac{1}{9}$ of 36 cents, or 4 cents, is $\frac{1}{100}$, and $\frac{100}{100}$ is 100 times 4 cents, which is 400 cents, or \$4; therefore, etc.

2. What principal, at 5 per cent., will gain 60 cents in 2 years?

3. What principal, at 8 per cent., will gain $2 in 6 months?

4. What principal, at 10 per cent., will gain $15 in 1 year?

5. What principal, at 8 per cent., will gain $60 in 3 years and 9 months?

6. What principal, at 10 per cent., will *amount* to $6.60 in 1 year?

Solution. The *amount* of $1 for 1 year, at 10 per cent., is $1.10; $1.10 in $6.60, 6 times; therefore, $6, at 10 per cent., will amount to $6.60 in 1 year.

Or, the interest of any sum for 1 year, at 10 per cent., is $\frac{1}{10}$ of the principal, and this added to the principal, $\frac{10}{10}$, makes the *amount* = $\frac{11}{10}$ of the principal; hence, $6.60 is $\frac{11}{10}$ of the principal, and $\frac{1}{11}$ of $6.60, or 60 cents, is $\frac{1}{10}$, and $\frac{10}{10}$, or the whole principal, is 10 times 60 cents = 600 cents, or $6; therefore, etc.

7. What principal, at 5 per cent., will amount to $22 in 2 years? To $44 in 2 years?

8. What principal, at 8 per cent., will amount to $58 in 2 years? To $84 in $1\frac{1}{2}$ years?

9. What principal, at 4 per cent., will amount to $66 in 5 years? To $280 in 10 years?

10. What principal, at $12\frac{1}{2}$ per cent., will amount to $250 in 2 years? To $121 in 3 years?

11. What principal, at $16\frac{2}{3}$ per cent., will amount to $42 in 1 year? To $80 in 2 years?

12. What principal, at 6 per cent., will amount to $10.60 in 1 year? To $21.80 in 1 year and 6 months?

13. What sum can I obtain at a bank for a note of $80, payable in 60 days?

Solution. The interest of $80 for 60 days is 80 cents, which, taken from $80, leaves $79.20; therefore, etc.

14. What sum can I obtain at a bank for a note of $48, payable in 45 days? What sum for a note of $64, payable in 75 days?

SECTION NINTH.

LESSON I.

1. BOUGHT 2 lb. of figs, at 20 cents per pound, and gave in payment a quarter of a dollar, a dime, and the rest in cents; how many cents did I give?

2. I have a rectangular field 20 rods long and 15 rods wide; what will it cost to build a wall around it, at $1.50 a rod?

3. I have a field containing 3 roods and 10 rods; what is it worth, at $2 per square rod?

4. What is the cost of a load of wood which measures 1 cord and 2 cord feet, at 75 cents per cord foot?

5. How many days from July 4th, at 9 o'clock in the morning, to Aug. 10th, at 9 o'clock in the evening?

6. What cost 3 score bushels of corn, at $\frac{3}{4}$ of a dollar a bushel?

7. If a ship sail 180 miles each day, how far does she sail in a week?

8. Bought a load of wood which measured 9 cord feet; $\frac{4}{9}$ of the load was pine, at $4 a cord, and the rest was oak, at 75 cents a cord foot; what was the cost of the load?

9. What cost $10\frac{1}{2}$ acres of land, at $20 per acre?

10. I have a square piece of land each side of which is 33 feet long; how many rods is it round this piece?

11. If it takes $16\frac{1}{2}$ yards of silk to make a dress, what will the silk for 2 dresses cost, at $1\frac{1}{4}$ dollars per yard?

12. I have a rectangular garden which is 12 rods long and 7 rods wide; how many square rods does it contain?

13. Mr. Smith has a rectangular garden, 6 rods wide, and containing 48 square rods; how long is it?

14. Mr. French has a rectangular garden 10 rods long and containing 70 square rods; how wide is it?

15. A man bought a share of the Boston and Maine Railroad stock for $100, and sold it for $103; what per cent. on the purchase money did he gain?

16. A grocer bought a box of sugar for $12, and sold it so as to gain 5 per cent.; how much did he gain?

17. A merchant bought a piece of silk for $75, and sold it so as to gain 10 per cent.; how much did he gain? What did he receive for the silk?

18. A shoedealer bought boots at $4 a pair, and sold them at $5; what per cent. did he gain? He bought others at $5, and sold them at $4; what per cent. did he lose?

19. A grocer bought molasses at 40c. per gallon; how shall he sell it to gain 25 per cent.?

20. A merchant bought goods for $50, but, they being damaged, he is willing to lose 20 per cent.; for what sum is he willing to sell them?

21. A laborer agreed to work 5 months for 60 dollars; how much did he receive for a month? How much for a week, allowing 4 weeks to the month? How many shillings a day, allowing 6 working-days to the week?

22. If a melon is worth 8 oranges, how many melons are 44 oranges worth?

23. A man sold 24 hens at the rate of 3 for a dollar; how many dollars did he receive?

24. If wine is worth 24 cents a pint, what is 1 gill worth? What are 3 gills worth?

25. At 80 cents a bushel, what is a peck of corn worth? What is a half peck worth?

26. Five boys bought a sled for 90 cents, and sold it for 75 cents; sharing the loss equally, what did each boy lose by the bargains?

27. A man paid $35 for wood, at $7 a cord; how many cords did he buy?

28. A man paid $63 for 9 cords of wood; what was the price per cord?

29. A hound is 36 rods behind a fox, and both are running in the same direction; but the hound gains upon the fox 3 rods in a minute; in how many minutes will the hound overtake the fox?

30. A fox is 48 rods from a hound, and running 80 rods per minute; but the hound is pursuing him at the rate of 83 rods per minute; how many minutes can the fox run before the hound overtakes him?

31. What will 9 quarts of milk cost, if 6 quarts cost 24 cents?

32. A cask containing 56 gallons, has a pipe which discharges 8 gallons per minute; in how many minutes will the cask be emptied?

33. There is an empty cask that will hold 84 gallons, and, by a pipe, 12 gallons run into it in a minute; in how many minutes will the cask be filled?

34. An empty cask, that will hold 48 gallons, receives 10 gallons per minute by one pipe, and discharges 6 gallons per minute by another pipe; how many gallons remain in the cask in one minute? In what time will the cask be filled?

35. A vessel which contains 63 gallons of water, discharges 12 gallons per minute by one pipe, and receives 9 gallons per minute by another pipe; in how many minutes will the vessel be emptied?

36. If 1 man can do a piece of work in 42 days, in how many days can 7 men do it? In how many days can 12 men do it?

37. If 1 man can do a piece of work in 55 days, how many men can do the same in 5 days?

38. A man can do a piece of work in 66 hours; in how many days can he do it if he works 11 hours each day?

39. A man can do a certain job in 56 hours; how many hours a day must he work to do it in 7 days?

40. If you wish to put 72 pounds of butter into 6 boxes, how many pounds will you put into each box?

41. If you put 84 pounds of butter into boxes that will hold 7 pounds apiece, how many boxes will it take?

42. A farmer bought 9 yards of cloth, at $4 a yard, and paid for it with apples, at $3 a barrel; how many barrels did it take?

43. Two boys are 45 rods apart, and both running in the same direction; the hindmost boy gains on the other 5 rods in a minute; in how many minutes will he overtake him?

44. A grocer bought 5 firkins of butter for $50, and sold them so as to gain $10; what did he pay for each firkin? What did he gain on each? What did he receive for each?

45. How many yards of cloth, at $5 a yard, may be bought for 10 reams of paper, at $2 a ream?

LESSON II.

1. WHAT number added to twice itself gives 15?

Solution. Any number plus twice the same number is 3 times that number; therefore, 15 is 3 times the required number, and $\frac{1}{3}$ of 15, viz., 5, is the number.

2. What number added to 3 times itself gives 32?

3. William and George together caught 35 fishes, and William caught 3 as often as George caught 2; how many did each boy catch?

4. Two times a certain number, added to 3 times the same number, make 35; what is the number?

5. A boy gave 35 nuts to two of his companions, giving 4 nuts to one of them as often as he gave 3 to the other; how many did he give to each?

6. What number added to $\frac{1}{2}$ of itself gives 15?

7. A boy being asked how old he was, replied, I am half as old as my sister, and the sum of our ages is 27 years; how old was he?

8. A father gave 64 cents to his two sons, giving $\frac{3}{5}$ as many to the younger as to the older; how many cents did he give to each?

9. A boy divided 28 peaches among 3 of his companions, giving twice as many to the first as to the second, and twice as many to the second as to the third; how many did he give to each?

10. A boy gave 42 peaches to 3 of his companions, giving half as many to the first as to the second, and half as many to the second as to the third; how many did he give to each?

11. A and B together have 45 marbles, and A has $\frac{2}{3}$ as many as B; how many has each?

12. A man being asked how many sheep he had, replied that if he had what he then had, and $\frac{1}{2}$ and $\frac{1}{3}$ as many more, he should have 55; how many had he?

13. A boy having 48 apples, kept 4 of them himself and gave the rest to Reuben, David, and Samuel, giving $\frac{1}{2}$ as many to David and $\frac{1}{3}$ as many to Samuel as to Reuben; how many did he give to Reuben?

14. A merchant bought a piece of cloth, at $6 a yard, and another piece of the same length, at $2 a yard, and afterward sold the whole, at $4 a yard; did he either gain or lose by the transactions?

15. A fruiterer employed two boys, Frank and Arthur, to sell apples for him. The dealer intrusted 60 apples to each boy, and told him he might sell them at the rate of 5 apples for 2 cents, and Frank sold his as instructed; but Arthur thought he might sell his more readily if he sorted them according to their quality; so he separated them, putting 30 apples into one part of his basket, and the other 30 into another part. He sold the better lot at the rate of 2 apples for 1 cent, and the poorer lot at the rate of 3 for 1 cent. Now, did the two boys receive equal sums for their entire stock of apples? If not, which was the more profitable to his employer? How much?

16. A boy was asked how many chickens he had, when he replied that if he had as many more, and $\frac{1}{2}$ and $\frac{1}{3}$ as many more and 2 chickens, he should have 70; how many had he?

17. A boy being asked his age, said that if $\frac{1}{2}$ and $\frac{1}{4}$ of his age and 10 years more were added to his age, the sum would be 3 times his age; how old was he?

18. A pole 48 feet long has one third as much in the mud as in the water, and twice as much in the air as in the mud and water together; how many feet were there in the water?

19. Suppose that A lends money at 6 per cent., interest payable in bank bills, and that B buys 6 per cent. government bonds, at 20 per cent. premium, which makes the best investment if the bonds are payable in paper in 20 years and the interest on the bonds payable in gold annually, gold being at 50 per cent. premium? Why?

20. Which is the more profitable, to lend money at 6 per cent., or to buy 5 per cent. government bonds, at 10 per cent. premium, the other conditions being the same as in example 19?

LESSON III.

1. DIVIDE 32 into two such parts that the less part shall have the same relation to the greater that 3 has to 5.

Solution. 3 and 5 are 8; therefore, if 32 is divided into 8 equal parts, 3 of those parts will be the less of the two required numbers, and 5 of them will be the greater. 3 times $\frac{1}{8}$ of 32 are 12, the less part, and 5 times $\frac{1}{8}$ of 32 are 20, the greater part.

NOTE. This is usually called *Ratio.* The ratio of 3 to 5 is $\frac{3}{5}$, and the ratio of 12 to 20 is $\frac{12}{20}$, which can be reduced to $\frac{3}{5}$.

2. Two boys bought a melon for 40 cents, and cut

it into 5 equal parts. One boy took 3 of the parts, and the other took 2; what ought each to pay?

3. Two men hire a pasture for $60. One man keeps 5 cows in the pasture through the summer, and the other keeps 7; how much ought each to pay?

4. Two boys engage to cut a pile of wood for $33. One boy cuts 6 cords while the other cuts 5; how should the pay be divided?

5. Two men engage in trade. One furnishes $7000 and the other $5000, and in a year they gain $2400; how should the profits be divided?

6. A, B, and C hire a pasture for $100. A pastures 3 cows, B 8, and C 9; how many dollars ought each to pay?

7. David, Samuel, and John buy 80 marbles for 20 cents; David takes 1 marble as often as Samuel takes 4, and John 5; how many does each take? What shall each pay?

8. A and B trade in company, furnishing money in the ratio of $4 to $5. They gain $2700; what is the share of each?

9. Two travelers, 54 miles apart, approach each other, one at the rate of 4 miles an hour, and the other at the rate of 5 miles an hour; how far will each travel before they meet?

10. A gentleman gave 28 pears to some children; to each girl 4, and to each boy 3. There were as many boys as girls; how many pears did he give to all the boys? To all the girls?

11. Sarah had 12 cents, and Nancy had 18 cents; they paid all their money for 10 oranges; how many oranges ought each to receive?

12. A, B, and C traded in company. A put in $2 as often as B put in $3, and as often as C put in $4. They gained $63; what was the share of each?

13. A man failing in business, has property valued at $900; but he owes to A $300, to B $400, and to C

$500; what part of his debts can he pay? How many dollars to each creditor?

14. Addie and Georgie together solved 60 examples, and Addie solved $\frac{7}{8}$ as many as Georgie; how many did each solve?

15. The sum of the ages of William and James is 36 years, and James is $\frac{4}{5}$ as old as William; what is the age of each?

16. Divide $24 between A and B, giving to A $1 as often as $\frac{1}{2}$ of a dollar to B.

17. Edward had a certain number of cents, and found $\frac{3}{5}$ as many more, when he had 32 cents; how many cents will he have left after spending $\frac{2}{3}$ of what he found?

18. Three men, A, B, and C, traded in company. A put in $2 as often as B put in $3, and C put in such a sum that he received $36 of the $66 gained in the year; what was the share of the gain received by A and B, and how many dollars did C furnish as often as A furnished $2?

LESSON IV.

1. A AND B traded in company. A put in $10 for 2 months, and B put in $8 for 3 months. They gain $88; what is each one's share of the gain?

Solution. A's $10 for 2 months are the same as $20 for 1 month, and B's 8 for 3 months are the same as $24 for 1 month; therefore, the total capital may be considered $44, and A's share is $\frac{20}{44}$ or $\frac{5}{11}$, and B's share is $\frac{24}{44}$ or $\frac{6}{11}$; $\frac{5}{11}$ of $88 is $40, A's share of the gain, and $\frac{6}{11}$ of $88 is $48, B's share.

2. A and B hire a pasture for $11. A pastures 2 horses for 5 weeks, and B pastures 3 horses for 4 weeks; what is each one's share of the expense?

3. Two men, A and B, hire a pasture for $10, and agree that 2 cows shall be reckoned as 1 horse. A

pastures 2 horses and 4 cows for 3 weeks, and B pastures 3 horses and 2 cows for 2 weeks; how many dollars shall each pay?

4. A, B, and C traded in company. A put in \$2 as often as B put in \$3, and as often as C put in \$5. B's money was in twice as long as C's, and A's twice as long as B's. They gained \$38; what was each one's share of the gain?

5. A and B trade in company. A puts in \$10 for 2 months, and B puts in \$9 for 3 months. They lose \$94; what is the loss of each?

6. A and B cut a field of grain for \$42. A works 12 days, and B works 6 days. What sum should each receive, if B can do as much in 2 days as A can do in 3 days? How much does A earn per day? How much B?

7. Two men, A and B, agree to do a piece of work for \$88. A employs 3 men 4 days, and B employs 5 men 2 days; what part of the pay shall each have? How many dollars?

8. A man divided \$40 between his 4 sons and 3 daughters, giving each son ½ as much as each daughter; what was the share of each?

9. Divide 15 peaches between B and C, so that B may have 3 peaches more than C.

10. The sum of 2 numbers is 21, and the greater exceeds the less by 5; what are the numbers?

11. A and B together have 40 cents, and A has 10 cents less than B; how many cents has each?

12. David bought twice as many pears as John, and after David had eaten 6 and John 4, they together had 26; how many pears had each remaining?

13. If 2 melons are worth 12 oranges, and 8 oranges are worth 24 apples, how many apples shall I give for 3 melons?

14. If 2 oxen are worth 3 cows, and 4 cows are worth 24 sheep, what is 1 ox worth, supposing a sheep to be worth \$7?

15. A hare is 7 of his own leaps before a hound, and takes 5 leaps while the hound takes 3, but 4 of the hound's leaps are equal to 7 of the hare's; how many leaps must the hound take to gain on the hare the length of one of the hare's leaps? How many leaps must the hound take to catch the hare?

Solution. Since 1 of the hound's leaps is $\frac{7}{4}$ of a leap of the hare, 3 of the hound's leaps are equal to $\frac{21}{4}$, or $5\frac{1}{4}$ of the hare's leaps; hence, in taking 3 leaps the hound gains $\frac{1}{4}$ of one leap of the hare, and therefore he must take 4 times 3 leaps, or 12 leaps, to gain 1 leap of the hare; and to gain 7 leaps, he must take 7 times 12 leaps, or 84 leaps.

16. Charlie is 10 of his own steps before John, and takes 6 steps while John takes 5, but 3 of John's steps are equal to 4 of Charlie's; how many steps must John take to gain the length of one of Charlie's steps? How many steps can Charlie take before John will catch him?

17. A lady wishing to buy a certain number of yards of silk for a dress, finds that if she pays a dollar a yard she will have 5 dollars left; but if she pays a dollar and a half per yard, she must get trusted for 2 dollars; how many yards does she wish to buy?

18. A man and his wife would eat a sack of flour in 15 days, and after living together for 6 days the woman alone would consume the remainder in 24 days; how many days would the whole sack last the woman? How many days the man?

19. A boy being asked how many hens he had, said that if he had half as many more and four hens and a half, he should have 3 dozen; how many hens had he?

20. A farmer, questioned with regard to the number of his sheep, said that if he had as many more, half as many more, and 2 sheep and a half, he should have 100; how many had he?

21. A boy engaged to live with a man 30 days, agree-

ing to receive 3 shillings a day for every day he worked, and to pay 2 shillings a day for his board every day he played. At the end of the 30 days he received 60 shillings; how many days did he work?

22. A fish's head weighs 5 pounds, his tail weighs as much as his head added to half the weight of his body, and his body weighs as much as his head and tail together; what is the weight of the fish?

23. A and B start from towns which are 48 miles apart, and travel toward each other until they meet, when it appears that $\frac{1}{3}$ of the distance A has traveled is equal to $\frac{3}{7}$ of the distance B has traveled; how far has each traveled?

24. A purse and its contents are worth 41 shillings, and $\frac{3}{4}$ of the value of the purse are equal to $\frac{5}{7}$ of the value of the contents; what is the value of the purse?

25. Seven times a certain number is 15 more than 4 times the same number; what is the number;

26. What number is that to which if its half and its fourth be added, the sum will be 35?

27. What sum of money can I obtain at a bank for a note of $36, payable in 90 days?

28. If a 5 cent loaf weighs 24 ounces when flour is worth $6 per barrel, what ought it to weigh when flour is worth $9 per barrel?

29. A man failing in business, is able to pay only 80 per cent. of his debts; how much will he pay on a debt of $30?

30. Bought $8\frac{5}{16}$ lb. of honey at one time, $3\frac{6}{23}$ lb. at another time, $10\frac{11}{16}$ lb. at another, and $6\frac{17}{23}$ lb. at another; how many pounds did I buy at the four times?

31. A, B, and C can do a certain piece of work in 4 days. A can do $\frac{1}{9}$ of it in 1 day, and B can do $\frac{1}{12}$ of it in 1 day; in how many days can C do the whole work?

32. A gentleman being asked the time of day, replied that $\frac{2}{3}$ of the time past from noon was equal to $\frac{4}{5}$ of the time to come before midnight; what was the time?

Once		**Twice**		**3 times**		**4 times**		**5 times**		**6 times**	
1 is	1	1 are	2	1 are	3	1 are	4	1 are	5	1 are	6
2	2	2	4	2	6	2	8	2	10	2	12
3	3	3	6	3	9	3	12	3	15	3	18
4	4	4	8	4	12	4	16	4	20	4	24
5	5	5	10	5	15	5	20	5	25	5	30
6	6	6	12	6	18	6	24	6	30	6	36
7	7	7	14	7	21	7	28	7	35	7	42
8	8	8	16	8	24	8	32	8	40	8	48
9	9	9	18	9	27	9	36	9	45	9	54
10	10	10	20	10	30	10	40	10	50	10	60
11	11	11	22	11	33	11	44	11	55	11	66
12	12	12	24	12	36	12	48	12	60	12	72

7 times		**8 times**		**9 times**		**10 times**		**11 times**		**12 times**	
1 are	7	1 are	8	1 are	9	1 are	10	1 are	11	1 are	12
2	14	2	16	2	18	2	20	2	22	2	24
3	21	3	24	3	27	3	30	3	33	3	36
4	28	4	32	4	36	4	40	4	44	4	48
5	35	5	40	5	45	5	50	5	55	5	60
6	42	6	48	6	54	6	60	6	66	6	72
7	49	7	56	7	63	7	70	7	77	7	84
8	56	8	64	8	72	8	80	8	88	8	96
9	63	9	72	9	81	9	90	9	99	9	108
10	70	10	80	10	90	10	100	10	110	10	120
11	77	11	88	11	99	11	110	11	121	11	132
12	84	12	96	12	108	12	120	12	132	12	144

13 times		**14 times**		**15 times**		**16 times**		**17 times**		**18 times**	
1 are	13	1 are	14	1 are	15	1 are	16	1 are	17	1 are	18
2	26	2	28	2	30	2	32	2	34	2	36
3	39	3	42	3	45	3	48	3	51	3	54
4	52	4	56	4	60	4	64	4	68	4	72
5	65	5	70	5	75	5	80	5	85	5	90
6	78	6	84	6	90	6	96	6	102	6	108
7	91	7	98	7	105	7	112	7	119	7	126
8	104	8	112	8	120	8	128	8	136	8	144
9	117	9	126	9	135	9	144	9	153	9	162
10	130	10	140	10	150	10	160	10	170	10	180
11	143	11	154	11	165	11	176	11	187	11	198
12	156	12	168	12	180	12	192	12	204	12	216

19 times		**20 times**		**21 times**		**22 times**		**23 times**		**24 times**	
1 are	19	1 are	20	1 are	21	1 are	22	1 are	23	1 are	24
2	38	2	40	2	42	2	44	2	46	2	48
3	57	3	60	3	63	3	66	3	69	3	72
4	76	4	80	4	84	4	88	4	92	4	96
5	95	5	100	5	105	5	110	5	115	5	120
6	114	6	120	6	126	6	132	6	138	6	144
7	133	7	140	7	147	7	154	7	161	7	168
8	152	8	160	8	168	8	176	8	184	8	192
9	171	9	180	9	189	9	198	9	207	9	216
10	190	10	200	10	210	10	220	10	230	10	240
11	209	11	220	11	231	11	242	11	253	11	264
12	228	12	240	12	252	12	264	12	276	12	288

13 times		**14 times**		**15 times**		**16 times**		**17 times**		**18 times**	
13 are	169	13 are	182	13 are	195	13 are	208	13 are	221	13 are	234
14	182	14	196	14	210	14	224	14	238	14	252
15	195	15	210	15	225	15	240	15	255	15	270
16	208	16	224	16	240	16	256	16	272	16	288
17	221	17	238	17	255	17	272	17	289	17	306
18	234	18	252	18	270	18	288	18	306	18	324
19	247	19	266	19	285	19	304	19	323	19	342
20	260	20	280	20	300	20	320	20	340	20	360
21	273	21	294	21	315	21	336	21	357	21	378
22	286	22	308	22	330	22	352	22	374	22	396
23	299	23	322	23	345	23	368	23	391	23	414
24	312	24	336	24	360	24	384	24	408	24	432

19 times		**20 times**		**21 times**		**22 times**		**23 times**		**24 times**	
13 are	247	13 are	260	13 are	273	13 are	286	13 are	299	13 are	312
14	266	14	280	14	294	14	308	14	322	14	336
15	285	15	300	15	315	15	330	15	345	15	360
16	304	16	320	16	336	16	352	16	368	16	384
17	323	17	340	17	357	17	374	17	391	17	408
18	342	18	360	18	378	18	396	18	414	18	432
19	361	19	380	19	399	19	418	19	437	19	456
20	380	20	400	20	420	20	440	20	460	20	480
21	399	21	420	21	441	21	462	21	483	21	504
22	418	22	440	22	462	22	484	22	506	22	528
23	437	23	460	23	483	23	506	23	529	23	552
24	456	24	480	24	504	24	528	24	552	24	576

EXERCISES IN ADDITION.

No. 1.	No. 2.	No. 3.	No. 4.	No. 5.
4 + 3	5 + 5	6 + 7	7 + 8	8 + 11
5 + 6	6 + 8	5 + 9	6 + 11	12 + 4
6 + 4	4 + 7	8 + 4	5 + 7	10 + 11
7 + 7	7 + 3	12 + 3	4 + 12	5 + 12
8 + 6	8 + 5	10 + 6	9 + 6	6 + 5
No. 6.	**No. 7.**	**No. 8.**	**No. 9.**	**No. 10.**
6 + 12	8 + 7	4 + 5	7 + 10	11 + 3
5 + 8	7 + 9	6 + 10	4 + 6	4 + 8
7 + 6	4 + 4	5 + 11	5 + 3	7 + 5
8 + 9	9 + 3	9 + 7	7 + 11	12 + 6
10 + 8	10 + 5	11 + 11	12 + 8	11 + 10
No. 11.	**No. 12.**	**No. 13.**	**No. 14.**	**No. 15.**
8 + 12	10 + 4	12 + 5	11 + 6	11 + 5
11 + 7	6 + 9	11 + 9	10 + 7	12 + 9
9 + 9	7 + 12	10 + 3	6 + 3	8 + 10
7 + 4	11 + 8	12 + 11	10 + 9	9 + 11
8 + 8	4 + 10	5 + 10	12 + 10	12 + 12

EXERCISES IN SUBTRACTION.

No. 16.	No. 17.	No. 18.	No. 19.	No. 20.
13 — 4	15 — 12	18 — 3	21 — 10	14 — 11
15 — 10	13 — 7	16 — 7	13 — 11	19 — 7
16 — 6	16 — 4	13 — 12	15 — 7	16 — 12
14 — 12	18 — 9	19 — 6	22 — 7	20 — 8
20 — 10	14 — 5	21 — 3	20 — 12	21 — 6
No. 21.	**No. 22.**	**No. 23.**	**No. 24.**	**No. 25.**
15 — 8	21 — 4	19 — 12	20 — 6	17 — 12
13 — 6	19 — 10	22 — 10	13 — 8	24 — 7
16 — 10	13 — 9	24 — 6	16 — 9	21 — 8
23 — 7	15 — 6	14 — 9	18 — 11	15 — 11
24 — 8	20 — 7	23 — 11	17 — 9	17 — 7
No. 26.	**No. 27.**	**No. 28.**	**No. 29.**	**No. 30.**
22 — 8	22 — 4	21 — 9	24 — 10	22 — 11
19 — 9	21 — 7	18 — 8	22 — 5	17 — 6
14 — 7	14 — 8	23 — 6	16 — 8	24 — 12
16 — 5	17 — 11	22 — 9	18 — 7	23 — 8
23 — 9	18 — 6	19 — 8	17 — 8	21 — 5

EXPLANATION.—These Tables are designed for a brief exercise at the beginning or close of the daily recitation, and should be dwelt upon until the pupil is familiar with the various combinations. Let the pupil recite the table assigned to him, as the signs indicate; thus, in No. 1, 4 and 3 are 7, 5 and 6 are 11, etc.; in No. 16, 13 less 4 are 9, etc.; in No. 31, 6 times 8 are 48, etc.; in No. 46, 4 in 36, 9 times, etc.

EXERCISES IN MULTIPLICATION.

No. 31.	No. 32.	No. 33.	No. 34.	No. 35.
6 × 8	4 × 5	8 × 2	5 × 7	8 × 9
5 × 4	6 × 6	9 × 4	3 × 8	5 × 5
9 × 3	3 × 12	7 × 6	6 × 12	10 × 8
7 × 2	5 × 9	10 × 3	2 × 11	6 × 4
8 × 5	7 × 11	8 × 7	7 × 8	3 × 7

No. 36.	No. 37.	No. 38.	No. 39.	No. 40.
7 × 9	9 × 12	11 × 12	5 × 10	11 × 6
4 × 8	4 × 7	12 × 11	3 × 3	9 × 2
6 × 5	3 × 1	6 × 9	12 × 5	1 × 1
8 × 8	6 × 11	4 × 10	9 × 7	12 × 3
10 × 7	5 × 8	9 × 8	11 × 5	6 × 7

No. 41.	No. 42.	No. 43.	No. 44.	No. 45.
12 × 9	8 × 11	12 × 12	8 × 3	11 × 11
3 × 10	4 × 9	9 × 5	12 × 7	8 × 6
8 × 4	12 × 8	7 × 12	3 × 11	9 × 11
11 × 3	10 × 11	9 × 6	7 × 10	8 × 12
4 × 4	7 × 5	11 × 9	5 × 12	12 × 6

EXERCISES IN DIVISION.

No. 46.	No. 47.	No. 48.	No. 49.	No. 50.
36 ÷ 4	25 ÷ 5	72 ÷ 9	36 ÷ 12	88 ÷ 11
56 ÷ 7	48 ÷ 12	80 ÷ 8	49 ÷ 7	84 ÷ 12
42 ÷ 6	45 ÷ 9	27 ÷ 9	108 ÷ 9	110 ÷ 10
55 ÷ 11	28 ÷ 4	120 ÷ 10	77 ÷ 11	99 ÷ 9
48 ÷ 8	100 ÷ 10	99 ÷ 11	18 ÷ 3	132 ÷ 12

No. 51.	No. 52.	No. 53.	No. 54.	No. 55.
81 ÷ 9	60 ÷ 10	144 ÷ 12	56 ÷ 8	121 ÷ 11
18 ÷ 6	84 ÷ 7	63 ÷ 9	35 ÷ 7	42 ÷ 7
20 ÷ 4	33 ÷ 11	1 ÷ 1	132 ÷ 11	96 ÷ 12
35 ÷ 5	54 ÷ 6	32 ÷ 8	60 ÷ 5	72 ÷ 6
88 ÷ 8	60 ÷ 12	66 ÷ 11	110 ÷ 11	54 ÷ 9

No. 56.	No. 57.	No. 58.	No. 59.	No. 60.
64 ÷ 8	33 ÷ 3	45 ÷ 5	48 ÷ 4	36 ÷ 6
48 ÷ 6	72 ÷ 12	72 ÷ 8	36 ÷ 9	40 ÷ 8
120 ÷ 12	96 ÷ 8	63 ÷ 7	22 ÷ 11	32 ÷ 4
44 ÷ 11	66 ÷ 6	28 ÷ 7	30 ÷ 6	30 ÷ 3
70 ÷ 7	21 ÷ 7	108 ÷ 12	27 ÷ 3	90 ÷ 9

The tables may be profitably used in many ways. The columns may be added upward or downward, or substituting + for —, ×, and ÷, add *across* the page. Substituting — for ÷, the table for *division* becomes a table for subtraction. By dividing the dividend of one example by the divisor of another, mixed numbers may be obtained; thus, in Nos. 46 and 47, say 4 in 25, 5 in 36, 7 in 48, etc.

WRITTEN ARITHMETIC.

---

1. On page 26 the pupil has learned the names and meaning of the ten Arabic figures, 0, 1, 2, 3, 4, 5, 6, 7, 8, 9; and on page 27 he has learned the names and uses of the signs, $, =, +, —, ×, ÷.

2. The learner has also doubtless observed that the *value* of a figure depends not only on the *figure itself*, but also on the *place* the figure occupies; for example, 4, standing alone, means *four units, or ones;* but when the 4 is made to occupy the second place, by putting 0 or any other figure on the right of it, it becomes *four tens;* thus, 40 equals four *tens*, or forty; 46 equals four tens plus six, that is, forty-six, etc. Again, when *two* figures are at the right of 4, the 4 becomes *four hundreds*, which equals ten times four tens; that is, each remove of a figure one place toward the left makes its value *ten times* as much as it was before. **This will be seen in the following Table:**

Hundreds of Thousands.	Tens of Thousands,	Thousands,	Hundreds	Tens,	Units,	
2	2	2,	2	2	2	= Two hundred and twenty-two thousand, two hundred and twenty-two.
			3	2	7	= Three hundred and twenty-seven.
		5,	4	9	6	= Five thousand four hundred and ninety-six.
	6	0,	2	4	3	= Sixty thousand, two hundred and forty-three.
6	0	1,	3	0	5	= Six hundred and one thousand, three hundred and five.

3. Let the pupil read the following numbers:

1. 56	5. 246	9. 3,742	13. 62,875
2. 78	6. 307	10. 5,897	14. 80,940
3. 87	7. 843	11. 8,609	15. 384,600
4. 95	8. 960	12. 9,054	16. 987,403

4. Write the following numbers in figures:

1. Forty-seven.
2. Sixty.
3. Two hundred and seventy-nine.
4. Five hundred and eight.
5. Three thousand, five hundred and sixty-two.
6. Eight thousand and two hundred.
7. Sixty-two thousand, five hundred and twenty.
8. One hundred and six thousand, two hundred and four.
9. Four hundred thousand, four hundred and thirteen.

ADDITION.

5. BY ADDITION we find how many units there are in two or more numbers taken together. The *result* of the addition is called the SUM, or AMOUNT.

6. To add when the amount of each column is less than ten.

Ex. 1. A farmer sold 125 bushels of corn, 342 bushels of oats, and 231 bushels of wheat; how many bushels of grain did he sell? *Ans.* 698.

OPERATION.

```
      125
      342
      231
     ----
Sum,  698
```

Having set the numbers so that units stand under units, tens under tens, etc., add the units; thus, 1 and 2 are 3, and 5 are 8, and set the 8 under the column of units. Then add the tens; thus, 3 and 4 are 7, and 2 are 9, and set the 9 under the column of tens, and so proceed till all the columns are added.

In like manner add the following examples:

2.	3.	4.	5.	6.	7.	8.
143	311	126	2604	324	124	2143
235	120	211	4132	241	371	124
421	247	242	1231	23	200	3010

9.	10.	11.	12.	13.	14.	15.
2101	1321	2144	2103	4162	203	2130
2234	2134	3241	412	1304	5120	3247
3120	3200	1302	3120	2013	62	1310

16. A man paid $112 for a horse, $235 for a chaise, and $41 for a harness; what did he pay for all?

17. A gardener has 231 apple trees, 124 pear trees, 322 peach trees, and 200 cherry trees; how many trees has he?

18. Add together 205, 342, 120, and 212.

19. Add together 134, 213, 401, and 141.

20. What is the sum of 1231, 3214, and 4323?

21. What is the sum of 5100, 1424, and 2432?

7. To add when the amount of any column is ten or more.

22. A man bought 4 farms. The first contained 345 acres, the second 244 acres, the third 346 acres, and the fourth 638 acres; how many acres were there in the 4 farms?

OPERATION.

```
      345
      244
      346
      638
Sum, 1573
```

We find the sum of the units to be 23, or 2 tens and 3 units. We write the 3 units under the column of units, and add the 2 tens to the tens in the next column, making 17 tens, or 1 hundred and 7 tens. The 7 tens are written in the place of tens, and the 1 hundred, added to the hundreds in the example, giving 15 hundreds, or 1 thousand and 5 hundreds; and when these figures are written in their respective places, the example is solved.

8. In the same manner add the numbers in the following *short columns*; also add *across* the page, as suggested by the *signs*.

23.	4236 +	3648 +	7264 +	8324 +	3663 +	2547
24.	2743	2575	8923	7087	8729	3826
25.	3254	1835	6845	3579	5273	9483
26.	2572	1247	3468	9200	6004	6245

27.	2356 +	1427 +	3240 +	5124 +	6312 +	2436
28.	5123	4172	3361	2534	1523	8240
29.	3245	1843	6024	3201	5430	5376
30.	3600	3412	5103	7120	6004	8241

31.	4235 +	1063 +	2714 +	3026 +	8124 +	3654
32.	6004	2130	184	24	381	298
33.	2146	401	8002	315	84	5423
34.	2401	7103	516	284	511	27
35.	1020	24	33	4123	871	4820

36.	36427 +	45326 +	52674 +	38956 +	2138
37.	54163	3402	254	4672	4694
38.	72412	68240	3529	388	5368
39.	43675	367	87006	54	2846
40.	12341	3412	61	4102	3714

41. A grain-dealer bought 375 bushels of wheat of A, 564 bushels of B, 2346 bushels of C, and 897 bushels of D; how many bushels of wheat did he buy?

42. I paid $3464 for a farm, $4875 for a house, $15625 for a mill, and $3375 for 30 shares of the Boston and Maine Railroad stock; how much did I pay for all this property?

43. How many are 63 + 79 + 144 + 28? *Ans.* 314.

44. How many are 469 + 8742 + 92748 + 869?

SUBTRACTION.

9. By Subtraction we take a less number from a greater number to find their difference.

The greater number is called the Minuend; the less number, the Subtrahend; the *difference*, the Remainder.

10. To subtract when no figure in the subtrahend is larger than the figure over it.

Ex. 1. From 748 take 236.

OPERATION.

Minuend,	748
Subtrahend,	236
Remainder,	512

Taking 6 from 8 leaves 2, 3 from 4 leaves 1, and 2 from 7 leaves 5; therefore, the remainder is 512.

	2.	3.	4.	5.	6.
From	7658	5703	8287	68295	84695
Take	3427	4301	3144	36241	32442
Ans.	4231				

	7.	8.	9.	10.	11.
From	87426	70827	3547	7643	35784
Take	45105	40326	2136	2431	13251

12. A farmer bought a farm for \$3625, and sold it for \$4986; how much did he gain? *Ans.* \$1361.

13. In 1860 the population of Maine was 628276, and that of New Hamsphire was 326072; how many more people were there in Maine than in New Hampshire?

11. To subtract when any figure in the subtrahend is greater than the figure over it.

14. From 982 take 137.

OPERATION.
Min. 982
Sub. 137
Rem. 845

As we cannot take 7 units from 2 units, *one* of the *tens* is put with the two units, making 12 units; and then 7 units from 12 units leave 5 units. Now as *one* of the 8 *tens* has been put with the 2 *units*, only 7 *tens* remain in the minuend, and 3 tens from 7 tens leave 4 tens; and, finally, 1 hundred from 9 hundreds leaves 8 hundreds; therefore, the entire remainder is 845.

12. In the same manner solve the following examples, taking each lower number from the one over it in each example; also subtract in the manner indicated by the *signs*.

15.	6379 — 3245	8268 — 7614	8078 — 5384
	2184 — 1528	3427 — 2817	2635 — 1748
16.	7683 — 3846	9463 — 4879	5009 — 2743
	5947 — 1072	8270 — 3948	2746 — 1834
17.	8725 — 3684	7000 — 4763	5948 — 3609
	6943 — 2948	3749 — 3628	4782 — 2948
18.	5402 — 3712	8760 — 2487	9046 — 8697
	4802 — 824	4837 — 948	5609 — 849
19.	6948 — 3879	5608 — 3742	5907 — 3764
	4062 — 2983	4924 — 936	3298 — 2643
20.	7829 — 6420	7869 — 3942	6308 — 5380
	943 — 784	3842 — 984	5709 — 648
21.	8037 — 4682	9293 — 6354	8725 — 3648
	5643 — 3486	8062 — 4630	4592 — 2000
22.	9999 — 6849	6849 — 5468	8234 — 6845
	7864 — 5986	3784 — 2407	3729 — 2698

23. Washington was born in 1732, and died in 1799; at what age did he die?

24. Methuselah died at the age of 969 years, and Washington at 67; what was the difference of their ages?

25. A merchant sold goods for $3427, and thereby gained $1224; what did the goods cost him?

26. In a certain battle 27345 men were engaged on one side, and 23598 on the other; how many more were engaged on one side than on the other?

27. The distance round the earth is about 24856 miles, and the distance through it is about 7912 miles; how much further round it than through it?

28. How many years have passed since the discovery of America in 1492?

MULTIPLICATION.

13. By Multiplication we find how many units there are in any number of times a given number.

The number to be repeated is called the Multiplicand; the number which shows how many times the multiplicand is to be taken is called the Multiplier; the *result* of the multiplication is called the Product. The *Multiplicand* and *Multiplier* are called Factors.

14. To multiply by a single figure.

Ex. 1. In 1 bushel there are 32 quarts; how many quarts are there in 9 bushels?

OPERATION.	
Multiplicand,	32
Multiplier,	9
Product,	288

Having set the factors as in the margin, we say, 9 times 2 units are 18 units = 1 ten and 8 units; write the 8 units in units' place, and then say, 9 times 3 tens are 27 tens, which, increased by the 1 ten previously obtained, make 28 tens = 2 hundreds and 8 tens, and these, written in the place of hundreds and tens respectively, give the true product, 288qt.

	2.	3.	4.	5.	6.
Multiplicand,	364	573	7243	436	3405
Multiplier,	3	4	5	6	7
Product,	1092	2292	36215	2616	23835

7.	8.	9.	10.	11.	12.
3259	2085	3706	8765	7432	5984
6	4	2	8	5	6

13. If 1 horse is worth $125, what are 7 horses worth?

14. In 1 year there are 365 days; how many days are there in 3 years?

15. If the President of the United States saves $5648 in 1 year, how much, at the same rate, will he save in 4 years?

16. How many yards of cloth are there in 9 bales, each bale containing 625 yards?

17. Multiply 356 by 7.
18. Multiply 874 by 3.
19. Multiply 368 by 5.
20. Multiply 759 by 4.
21. 3742 × 8 = ?
22. 6536 × 6 = ?
23. 3042 × 2 = ?
24. 7608 × 9 = ?

15. To multiply by two or more figures.

25. How many quarts are there in 49 bushels?

OPERATION.

Multiplicand,	32
Multiplier,	49
Partial Products,	288
	128
True product,	1568

First multiply by 9, as though 9 were the only figure in the multiplier; then multiply by 4, and set the first figure of this product under the 4; finally, add the partial products together, and the sum will be the true product.

16. In the same manner solve the following examples, multiplying each upper number by the one under it in each example; also multiply in the manner indicated by the *signs*.

26.	345 × 24	504 × 32	7063 × 25
	37 × 56	56 × 45	62 × 74
27.	432 × 243	524 × 325	540 × 367
	724 × 36	464 × 426	842 × 304
28.	4326 × 28	6304 × 56	3068 × 384
	325 × 32	542 × 72	247 × 265
29.	5648 × 59	4008 × 63	5308 × 872
	364 × 38	3604 × 75	2046 × 635
30.	3564 × 48	9604 × 49	6350 × 358
	2435 × 84	376 × 84	4635 × 496
31.	5642 × 67	4768 × 95	3465 × 386
	325 × 84	347 × 46	7308 × 947

32. If a steamboat sails 288 miles per day, how far will she sail in 27 days?

DIVISION.

17. By Division we find how many times one number is contained in another.

The number to be divided, is called the Dividend; the number by which to divide is called the Divisor; the number of times the dividend contains the divisor is called the Quotient. If anything is left after dividing, it is called the Remainder.

18. To perform Short Division.

Ex. 1. How many weeks are there in 364 days?

OPERATION.

Divisor, 7)364 Dividend.
Quotient, 52

In dividing we first say, 7 in 36, 5 times and 1 remainder; set the quotient, 5, under the 6 of the dividend, and then, *imagining* the remainder, 1, placed *before* the 4, say, 7 in 14, 2 times; set the 2 under the 4, and thus we find the quotient, 52. This process is called *Short Division.*

2.	3.	4.	5.
Divisor, 6)162 Dividend.	5)875	8)432	6)456
Quotient, 27			

6.	7.	8.	9.	10.
7)4417	8)4504	2)1764	3)2847	5)3465

11.	12.	13.	14.	15.
4)2296	9)2916	7)3122	8)4320	6)3816

16. How much sugar, at 9 cents a pound, can be bought for 342 cents, or $3.42? *Ans.* 38lb.

17. How many pigs, at $5 each, can be bought for $285?

18. If a horse travels 7 miles per hour, in how many hours will he travel 1001 miles?

19. Divide 161 by 7.	24. 9732 ÷ 4 = ?
20. Divide 204 by 6.	25. 1768 ÷ 2 = ?
21. Divide 475 by 5.	26. 7648 ÷ 8 = ?
22. Divide 345 by 3.	27. 4428 ÷ 6 = ?
23. Divide 846 by 9.	28. 3762 ÷ 9 = ?

19. To perform Long Division.

Ex. 29. Paid \$4932 for 9 acres of land; what was the price per acre?

OPERATION.

9)4932(548
45
43
36
72
72
0

Having set the divisor and dividend as in Short Division, draw a curve at the right of the dividend, and then say, 9 in 49, 5 times, and set the 5 at the right of the dividend. Then multiply the divisior by the quotient, 5, and set the product, 45, under the 49 of the dividend, and substract the 45 from the 49. To the remainder, 4, annex 3, the next figure of the dividend, so forming a new partial dividend, and then say, 9 in 43, 4 times, and set the 4 as the next figure of the quotient. Multiply the divisor by this new quotient-figure, and substract the product from the partial dividend. Proceed in this manner until the whole dividend has been divided. This is called *Long Division.*

20. In the same manner solve the following examples; also divide in the manner indicated by the *signs.*

30.	31.	32.
8)3568(446	12)6564(547	13)4753(365
32	60	39
36	56	85
32	48	78
48	84	73
48	84	65
0	0	Remainder, 8

33.	Divide	1728 ÷ 72	90090 ÷ 195
	By	144 ÷ 12	4095 ÷ 15
34.	Divide	309120 ÷ 805	351900 ÷ 391
	By	25760 ÷ 23	9775 ÷ 23
35.	Divide	46385 ÷ 563	12463 ÷ 816
	By	2435 ÷ 42	3746 ÷ 24
36.	Divide	54687 ÷ 348	27462 ÷ 948
	By	2731 ÷ 16	3724 ÷ 39

37. How much tea, at 35 cents per pound, can be bought for \$8.75?

38. How much flour, at $9 per barrel, can be bought for $3762?

39. In how many days can a man walk 252 miles, if he walks 21 miles per day?

40. How many days are there in 1728 hours?

41. How many dollars in 3168 shillings, if 6 shillings make a dollar? If 8 shillings make a dollar?

42. In how many days will a ship sail 5040 miles, if she sails 144 miles per day?

43. A drover paid $1431 for 27 oxen; what was the average price per ox?

44. In how many hours will a locomotive run 1225 miles, if it runs 25 miles per hour?

45. A farmer raised 1458 bushels of corn on 27 acres; how many bushels per acre did he raise?

REDUCTION.

21. From the Lessons in Section V, the learner will readily see the nature of the following examples. The changing or reducing numbers from one name or denomination to another, without altering their values, is called REDUCTION.

1. In 3bush. 2pk. 5qt. 1pt. how many pints?

OPERATION.

bush.	pk.	qt.	pt.
3	2	5	1

4
14 pk.
8
117 qt.
2
235 pt., *Ans.*

Multiply the 3 by 4 to change the bushels to pecks; to the product, 12, add the 2pk. given in the example, and the result is 14pk.; then multiply the 14 by 8 to change the pecks to quarts; to the product, 112, add the 5qt. in the example, and the result is 117qt.; so proceed till the example is solved.

2. In 5gal. 3qt. 1pt. 2gi. how many gills?
3. In 3£ 4s. 9d. 3qr. how many farthings?
4. In 3wk. 4d. 6h. 12m. 20sec. how many seconds?
5. In 4yd. 2ft. 7in. how many inches?
6. Reduce 7yd. 3qr. 2na. to nails.
7. Reduce 3sq. m. 320a. 2r. 20sq. rd. to rods.
8. Reduce 3c. 5c. ft. 12cu. ft. 1654c. in. to inches.

9. Change 1843qr. to pounds, shillings, etc.

OPERATION.

```
4)1843qr.
12)460d. + 3qr.
 20)38s. + 4d.
    1£ + 18s.
Ans. 1£ 18s. 4d. 3qr.
```

First divide by 4 to reduce the farthings to pence, giving 460d. and 3qr.; then divide the 460 by 12 to reduce pence to shillings, giving 38s. and 4d.; then divide the 38 by 20 to reduce shillings to pounds, and thus obtain the *Ans.*

10. Change 387na. to yards, quarters, etc.
11. Change 16879gr. Troy Weight to pounds, ounces, etc.
12. Change 16879gr. Apothecaries' Weight to pounds, etc.
13. Change 716893dr. to tons, etc.
14. Reduce 5327rd. to miles, furlongs, etc.
15. Reduce 47386sq. in. to square yards, etc.
16. Reduce 356482c. in. to cubic yards, etc.
17. Reduce 876gi. to gallons, quarts, etc.
18. Reduce 647pt. to bushels, pecks, etc.
19. Reduce 753986sec. to weeks, days, etc.
20. At 5c. a gill, what will 3gal. 2qt. 1pt. 3gi. of wine cost?
21. At 3c. apiece, what will 3 gross, 5 dozen, and 6 buttons cost?
22. At 2c. a sheet, what will 3 reams, 6 quires, and 8 sheets of paper cost?
23. At $16 per ounce, what are 3 lb. 7oz. of gold worth?
24. At 2c. per ounce, what are 15 lb. 14oz. of iron worth?
25. At 2c. a pint, what are 5bush. 3pk. 6qt. 1pt. of corn worth?
26. If 1oz. of iron will make 3 nails, how many nails may be made of 7 lb. 15oz. of iron?
27. At 2c. a pint, what are 5gal. 2qt. 1pt. of milk worth?
28. At 3c. apiece, what are 6 dozen and 9 oranges worth?
29. At 2c. apiece, what are 3 score and 15 lemons worth?

MISCELLANEOUS EXAMPLES.

1. Mr. Stone had 364 acres of land in one piece, and 274 acres in another piece; but he has sold 125 acres from the first, and 94 acres from the other; how many acres has he now?

2. Mr. Pray paid $115 for one piece of carpet, and $112 for another; he sold both pieces for $250; how much did he gain?

3. What is the value of 65 acres of land at $137 per acre?

4. How many men are there in a regiment of 15 companies, having 133 men in each company?

5. What will 36 barrels of flour cost, if 9 barrels cost $99?

6. What will 13 acres of land cost, if 52 acres cost $7488?

7. If 12 men can cut 60 cords of wood in a week, how many cords can 17 men cut in the same time?

8. How many rods are there in 875 miles, 5 furlongs, and 33 rods?

9. How many farthings are there in 354£ 17s. 9d. 1qr.?

10. A man has two farms which together contain 432 acres, and one farm is 7 times as large as the other; how many acres are there in each?

11. Divide 1728 into two such parts that the first shall be 11 times as large as the second.

12. Reduce 940819 drams to tons, etc.

13. Reduce 907336 seconds to weeks, days, etc.

14. If a man can walk 23 miles per day, in how many days can he walk 391 miles?

15. A road-builder employed 33 men, giving the same wages to each, and at the end of 2 months it took $1584 to pay them; what were the wages of each per month?

16. In a certain house there are 8 rooms, having 3 windows in each room, and 12 panes of glass in each window; how many panes of glass are there in the house?

17. A merchant having 3 pieces of cloth measuring 63yd., 45yd., and 56yd. severally, sold 22yd. from the first, 31yd. from the second, and 14yd. from the third; how many yards had he remaining?

18. Mr. Holt owed $3462, but paid $1362 in May, and $897 in June; how much did he still owe?

19. How many are $876 + 392 + 648 - 987$?

20. How many are $689 + 9642 + 87 - 398$?

21. From the sum of 384 and 426, take the difference between 567 and 432.

22. How many are 876×43? 968×382?

23. How many are $1584 \div 66$? $21364 \div 763$?

24. How many are $38 \times 46 - 24 \times 26$?

25. How many are $87 \times 33 + 36 \times 27$?

26. What cost 54 tons of hay, at $18 per ton?

27. A farmer sold 75 bushels of wheat, at $2 per bushel, for which he received 15 yards of cloth, at $4 per yard, and the balance in money; how much money did he receive?

28. A man's income is $1687 a year, and his expenses are $4 a day; what does he save in a year of 365 days?

29. The earth, in going round the sun, moves 19 miles per second; how far does it move in an hour?

30. Light moves 192000 miles in a second; how far does it move in an hour?

31. I owe 3 notes, whose sum is $368. One of these notes is for $125, another for $84; for what is the third one?

32. A and B start from the same place, and travel in the same direction, A at the rate of 48 miles, and B 64 mile per day; how far apart are they in 1 day? In 24 days?

33. A and B start from the same place, and travel in opposite directions, A at the rate of 36 miles, and B 44 miles, per day; how far apart are they in 35 days?

34. The President of the United States receives a salary of $25000 a year; what will he save in a year of 365 days, if his expenses are $40 per day?

35. In 1 hogshead of wine there are 63 gallons; how many gallons in 36 hogsheads?

36. How many yards of cloth in 25 bales, each bale containing 54 pieces, and each piece 36 yards?

37. The last transit of Venus across the sun's face was in 1769, and the next will be in 1874; how many years between these two transits?

38. Bonaparte was born in the year 1769, and lived 52 years; in what year did he die?

39. The first settlement in New England was made at Plymouth in 1620; how many years is it since that time?

40. How long since the Declaration of American Independence in 1776?

41. A rectangular piece of ground is 72 rods long and 36 rods wide; how many square rods are there in the piece? How many rods is it round the piece?

42. A piece of ground is 42 rods square; what will it cost to build a wall around it, at $2 per rod? What is the piece of ground worth, at $3 per square rod?